VIPER ALLEY
A Combat History of the F-16

Mike Guardia

Copyright 2025 © Mike Guardia

From *Debrief: A Complete History of US Aerial Engagements, 1981-present* by Craig Brown. Used by permission of Schiffer Publishing. Any third-party use of this material, outside of this publication, is prohibited. Interested parties must apply directly to Schiffer Publishing for permission.

From *F-16 Fighting Falcon Units of Operation Iraqi Freedom* by Steve Davies and Doug Dildy, Osprey Publishing, used in accordance with UK copyright laws regarding Fair Usage, quoting 800 words or less.

From *Vipers in the Storm: Diary of a Gulf War Fighter Pilot* by Keith Rosenkranz. Used by permission of McGraw-Hill. Any third-party use of this material, outside of this publication, is prohibited. Interested parties must apply directly to McGraw-Hill for permission.

From *Operation Allied Force - Volume 1: Air War Over Serbia, 1999*, by Bojan Dimitrijevic and Jovica Draganić, Helios & Company, used in accordance with UK copyright laws regarding Fair Usage, quoting 800 words or less.

From *War in Ukraine - Volume 6: The Air War February-March 2022*, by Tom Cooper, Adrien Fontanellaz, and Milos Sipos; Helios & Company, used in accordance with UK copyright laws regarding Fair Usage, quoting 800 words or less.

Published by Magnum Books
PO Box 1661
Maple Grove, MN 55311

www.mikeguardia.com

ISBN-13: 979-8-9917981-4-3

All rights reserved, including the right to reproduce this book or any part of this book in any form by any means including digitized forms that can be encoded, stored, and retrieved from any media including computer disks, CD-ROM, computer databases, and network servers without permission of the publisher except for quotes and small passages included in book reviews. Copies of this book may be purchased for educational, business or promotional use.

Also by Mike Guardia

The Combat Diaries
Skybreak
Tomcat Fury
Wings of Fire
Foxbat Tales

An F-16 from the 555th Fighter Squadron (Aviano Air Base, Italy) in flight over Afghanistan in support of Operation Enduring Freedom, March 2011. *US Air Force*

An F-16 from Shaw Air Force Base during Sentry Savannah 2016, the Air National Guard's premier counter-air exercise for 4th- and 5th-Generation fighters. *US Air Force*

Table of Contents

Introduction	1
Chapter 1: Bird of Prey	5
Chapter 2: Wings over the Negev	23
Chapter 3: Desert Talons	81
Photo Section	143
Chapter 4: Grey Zone Falcons	157
Chapter 5: This Kind of War	213
Chapter 6: The Second Storm	255
Chapter 7: Arab Winter	297
Chapter 8: Brushfires	339
Epilogue: The Viper's Enduring Legacy	379
Select Bibliography	385

Artist David Poole created this 1992 rendition of a flight of F-16s from the 169th Tactical Fighter Wing (South Carolina Air National Guard), during the Allied air offensive of Operation Desert Storm, 1991. *US Air Force*

Introduction

The F-16 Fighting Falcon has earned a reputation that few aircraft will ever match. Affectionately known as the "Viper" to its pilots, the F-16 has become a legend of modern-day air power. From the hot skies of the Middle East to the grey zone intercepts over Yugoslavia, the F-16 has flown more missions, fought in more conflicts, and served under more flags than any other Western fighter of its generation.

Born from the lessons learned in Vietnam, the F-16 was initially designed as a lightweight, low-cost alternative to the F-15 Eagle. What emerged, however, was a highly-adaptable strike fighter with capabilities far beyond what its Cold War designers had envisioned. Indeed, the F-16 wasn't just a new fighter, it was a revolution in aircraft design. With more than 4,600 units built (and millions of flight hours logged), it remains one of the most versatile and battle-tested aircraft of the modern age.

Following its operational debut in 1979, the Viper has been tested in combat zones far and wide. From its first dogfights over Lebanon to its latter-day service in Ukraine, the F-16 has proven itself time and again. Whether flying intercepts over the Persian Gulf or targeting Taliban strongholds in Afghanistan, the Viper has played a role in every major conflict since the end of the Vietnam War.

Viper Alley is the definitive story of these harrowing missions. This book is a *combat* history: An unapologetic, unbiased, no-holds-barred account of the F-16 in battle. Drawing from de-classified records, pilot interviews, and after-action reports, *Viper Alley* traces the F-16's journey through four decades of aerial combat.

These stories, however, are not without their darker moments.

No aircraft is invincible; and the Viper is no exception. Shootdowns, crashes, collisions, and even fratricide have all been part of the F-16's journey. The reader must bear in mind, however, that these incidents are discussed not for the sake of sensationalizing, but to present an earnest portrayal of the risks inherent to aerial warfare.

The reader will also find a critical assessment of how the F-16 has performed against a growing number of peer and near-peer threats. Despite the advent of 5th-generation adversaries like the J-20 and Su-57, the Viper has continued to evolve and retain its lethality on the modern battlefield. In fact, the most-recent versions of the F-16 are expected to remain in service well into the 2040s.

This book discusses the F-16's operational history both chronologically *and* thematically, tracing the plane from its first combat missions in Israeli service to its role in present-day conflicts. Each chapter focuses on a particular campaign or era. These include: Israeli F-16 operations in Lebanon, Iraq, and over the Gaza Strip; NATO F-16s in the Balkans; American F-16s during Operation Desert Storm and the ensuing No-Fly Zone missions; the mass mobilizations in the Global War on Terror; and the more recent missions over Syria and Ukraine. These chapters also discuss the F-16's lesser-known conflicts, including the Indo-Pakistani air wars, Turkish F-16s in combat over northern Syria, and the UAE's Viper deployments in the Yemeni Civil War. The international scope of these stories demonstrates both the reality of the F-16's widespread service and its continued relevance in the realm of modern warfare.

At its heart, however, the F-16's story is one of

resilience, adaptability, and surpassing expectations. It's the story of how a lightweight prototype built for B-list combat roles became the backbone of NATO's air power. It's the story of how a seemingly bargain-basement aircraft (designed as an understudy to the F-15) became the most prolific fighter jet in the world. It's the story of how a fighter once dismissed as a niche aircraft earned the loyalty of its pilots and a begrudging respect from its enemies.

Viper Alley pays tribute not just to the aircraft itself, but to those who built it, the tireless crews who maintained it, and the pilots who flew it into harm's way. The F-16 may be a machine, but its legacy is profoundly *human*. As it continues to fly well into the 21st Century, the F-16 Fighting Falcon remains an enduring symbol of modern airpower.

An F-16 from the 79th Fighter Squadron during a midair refuel at Exercise Red Flag over the Nevada Test and Training Range, 2016. *US Air Force*

An F-16 from the 180th Fighter Wing (Ohio Air National Guard) in the early morning hours before a training exercise at MacDill AFB, February 2017. The anticipated maneuvers were against CF-18 Hornets from the Royal Canadian Air Force. *US Air Force*

Chapter 1:
Bird of Prey

When freedom is on the line, the F-16 answers the call. The Fighting Falcon delivers peace through strength, keeping America and its allies ahead of every threat. – Lockheed Martin

In the smoldering aftermath of Vietnam, the US Air Force found itself at a critical juncture. Eight years of aerial combat had exposed some harsh realities about the current state of tactical air doctrines. Heavily armed and highly sophisticated aircraft like the F-4 Phantom II had been designed under Cold War assumptions that aerial dogfights were obsolete, and beyond-visual-range (BVR) missile encounters would become the new norm.

Yet, in the skies over Laos and North Vietnam, planes like the MiG-21 proved otherwise. American pilots were often forced into tight-turning battles against lighter, leaner, and more maneuverable aircraft. And, without their own built-in guns, the early Phantoms were often helpless once their missiles had been depleted.

By the end of the war, there could be little debate: The current fleet of American fighter jets lacked the capacity for high-speed aerial combat. Begrudgingly, the Air Force realized that raw performance metrics—speed, altitude, and missile range—were only part of the equation. What truly mattered in a dogfight was *energy maneuverability*: the ability of an aircraft to maintain its momentum during a fight, turn on a dime, and outlast its opponent.

This chapter explores the pivotal phases in the

creation, evolution, and eventual deployment of the F-16. It explores how a handful of revolutionary thinkers developed a lighter and more agile aircraft, breaking away from traditional bulkier designs like the F-100 series and F-4 Phantom II. This journey unfolds through the eyes of those who dared to defy established norms, setting new standards for conventional air power. These are the transformative milestones that defined the F-16's path from concept to combat readiness.

In many ways, the shift in America's air combat doctrine began with the relentless fervor of Colonel John Boyd (USAF) and defense analyst Thomas Christie. Boyd was a highly-decorated fighter pilot, known as "40-Second Boyd" for his legendary ability to defeat *any* adversary in a simulated dogfight in under 40 seconds. His other nicknames included "Genghis John," for his quick temper and confrontational personality; the "Mad Major" for the passion of his convictions; and the "Ghetto Colonel" for his spartan lifestyle.

But Boyd wasn't just a fighter pilot; he was a strategic thinker who analyzed dogfighting with an intensity that sparked a revolution. His philosophy challenged conventional thinking, arguing that speed and maneuverability were just as critical as high-tech weaponry. Thomas Christie, supporting Boyd's vision with a mathematical rigor, provided compelling data that reinforced this need for change. Together, Boyd and Christie developed the Energy-Maneuverability (EM) Theory—a framework for evaluating an aircraft's ability to gain *and* maintain energy in a dogfight, the essence of air combat agility.

EM Theory proved that many of the Air Force's top-dollar aircraft were highly flawed. Boyd argued that the service had become obsessed with airspeed and long-distance penetration at the expense of agility, responsiveness, and survivability in a dogfight. His findings challenged decades of air power orthodoxy (and ran afoul of many high-ranking officials), but Boyd was undeterred. With a small cadre of like-minded reformers, later dubbed the "Fighter Mafia," John Boyd began a quiet insurgency within the halls of the Pentagon.

Ironically, postwar budget constraints, combined with the brutal lessons from Vietnam, created just the right opportunity for Boyd's ideas to reach fertile ground. In 1972, a modest initiative known as the Lightweight Fighter Program (LWF) began within the Air Force's Prototype Program Office. Its goal was to build a small, cheap, lightweight air superiority fighter with unparalleled maneuverability and acceleration.

It was a radical departure from the Air Force's trend towards increasingly expensive, multi-role platforms like the F-111. In short, LWF sought to prove that smaller, more-focused fighters could outperform their heavier cousins in real-world combat scenarios. But even with these mission parameters, LWF was never intended to be a procurement program. Rather, it was a technology demonstrator—a way to test and validate concepts without committing to a full production line.

Nevertheless, General Dynamics and the Northrop Corporation sensed an opportunity, respectively submitting prototypes for what would become the YF-16 and YF-17. The General Dynamics YF-16, designed under the leadership of Harry Hillaker, embodied Boyd's EM principles to the core. With its frameless bubble canopy, a reclined pilot seat to endure high-G forces, a fly-by-wire flight control system (the first of its kind in a fighter jet), and an emphasis on thrust-to-weight ratio, the YF-16 was a plane built for the *pilot*, not the bureaucracy. Northrop's YF-17, meanwhile, was equally impressive, with a powerful twin-engine configuration. The YF-17 was slightly heavier than its General Dynamics counterpart, but it boasted greater power. Still, each prototype embodied the very principles for which Boyd had campaigned: High thrust-to-weight ratios, low-drag profiles, and superior agility.

By early 1974, both prototypes were ready for testing. Though both planes were impressive, the YF-16 ultimately proved superior in terms of agility, acceleration, and cost. While the Air Force selected the YF-16 as its LWF standard-bearer, the YF-17 would go on to become the Navy's F/A-18 Hornet.

As the YF-16 began its journey from demonstrator to operational fighter, it was clear that John Boyd's strategies had done more than influence a single project. They had realigned the defining principles of air superiority. His ability to disrupt traditional thought patterns and introduce a more iterative, adaptable approach to aircraft design resonated far beyond LWF, influencing future generations of combat aircraft. This reform was not merely technical; it was *cultural*. Indeed, it challenged the settled norms about what a fighter should be and what it could achieve. It enabled a progressive shift in how military and aerospace industry leaders approached fighter aircraft, favoring adaptability and tactical advantage over brute force.

However, for the YF-16, winning the LWF competition was only the beginning. The Air Force quickly expanded the project into the Air Combat Fighter (ACF) competition, seeking to replace older aircraft like the F-104 and F-5. According to then-Secretary of Defense James R. Schlesinger, the ACF would not evaluate a pure fighter, but a *multirole* aircraft to complement the forthcoming F-15 Eagle. Alongside the YF-16, ACF would also evaluate a number of foreign designs, including Dassault's prototype Mirage F1M-53, the SEPECAT Jaguar, and Saab's experimental 37E. Northrop, meanwhile, submitted the P-530 Cobra, similar to the latter-day YF-17.

The ACF competition represented the ultimate battle of engineering and design prowess among the aerospace giants involved. Its core objective was to select a lightweight, cost-effective fighter that could meet the sophisticated demands of air supremacy. With an emphasis on agility, versatility, and economy, ACF wanted a fighter that was capable of sustaining high-performance levels without breaking the post-Vietnam budgetary constraints.

Testing, therefore, became an integral part of the ACF stakes. Various trials were organized, placing each prototype in scenarios that simulated the dynamics of real-world combat. Pilots, engineers, and analysts tirelessly collected data, observing everything from aerodynamic stability to radar cross-sections and avionics integration.

During these stakes, the YF-16 emerged as an early frontrunner primarily due to its groundbreaking innovations. It offered advanced aerodynamics and a fly-by-wire control system to enhance its maneuverability, thus giving the pilot an unprecedented level of responsiveness. More importantly, this enhancement didn't come at the expense of the pilot or the mission. Rather, it increased the plane's operational effectiveness while maintaining stability under stressful conditions.

These technological advancements epitomized what the Air Force sought: A fighter versatile enough to excel in dogfights *and* multirole operations. Although computer analyses predicted that ACF would be a close contest, the YF-16 handily won the competition, becoming the unanimous choice of the test pilots involved. The YF-16's innovative inlet design minimized drag while maximizing thrust. In fact, compared to its ACF competitors, the YF-16 held the edge across every critical performance metric. Its design reflected a commitment to innovation while understanding the practical demands of modern-day aerial combat.

On January 13, 1975, Secretary of the Air Force John L. McLucas announced the YF-16 as the winner of the ACF competition. Citing lower operational costs, greater range, and superior maneuverability (especially at supersonic speeds), McLucas further announced that the Air Force planned to order at least 650 full-scale development (FSD) F-16s by the end of the decade.

The first production order came in 1975, totaling eight FSD aircraft (six F-16A single-seaters and two F-16B two-seaters). For these production-model variants, General Dynamics lengthened the fuselage by 10.6 inches and fitted the airframe with a larger radome to accommodate the AN/APG-66 radar. The first production-run variants also featured a larger wing surface area, smaller tailfin, and enlarged ventral fins. Taken together, these modifications increased the F-16's weight by nearly 25% over the YF-16 prototype.

The initial FSD models were manufactured in Texas at General Dynamics' Fort Worth Division. The first production-standard F-16A took its maiden flight on August 7, 1978 and was accepted by the US Air Force on January 6, 1979. The F-16 officially received its name "Fighting Falcon" on July 21, 1980. Yet, almost from the start, its pilots and crews gave it the nickname "Viper," due to its perceived resemblance to a viper snake, as well as the fictional "Colonial Viper" starfighter from the popular television show *Battlestar Galactica*. The aircraft officially entered operational service with the 34th Fighter Squadron at Hill AFB, Utah in October 1980.

In its definitive form, the F-16 features a blended wing-body design, with smooth transitions between the fuselage and wing roots. This configuration minimizes drag and enhances lift. The mid-mounted, cropped-delta wing has a 40-degree sweep with a thickness-to-chord ratio of 4%, which offers excellent maneuverability at high angles of attack. With a wingspan of approximately 32.8 feet and an overall length of 49.3 feet, the Viper maintains a relatively small footprint, making it ideal for expeditionary warfare.

The single-piece, bubble canopy was another hallmark of the F-16's design. Providing an unobstructed, 360-degree view of the battlespace, the canopy design offers a critical advantage in dogfights where visual acquisition is still paramount. In fact, survivability and situational awareness were the guiding principles behind the F-16's cockpit design. It reflects a "man-machine interface" that was well ahead of its time, and has since become an industry standard. Within the cockpit itself, the pilot sits atop an ejection seat reclined to 30 degrees (unique among NATO fighters), which helps the pilot withstand higher G-forces by reducing effective gravitational loads on the body. This ergonomic design greatly enhances the pilot's physical endurance during air combat maneuvering.

Like most modern fighter jets, the cockpit includes a Head-Up Display (HUD) that projects critical flight data and targeting information onto the windshield, allowing the pilot to remain "heads-up" during combat. Later versions of the F-16 incorporated Multi-Function Displays (MFDs) and eventually featured a fully-digital glass cockpit. However, one of the most distinctive features across all versions of the Falcon is the side-stick controller, mounted on the right console rather than between the pilot's legs. This design allows for better control during high-G maneuvers and frees up space in the cockpit. Additionally, the Hands-On Throttle and Stick (HOTAS) configuration reduces pilot workload by placing all essential controls within finger reach.

Beneath the cockpit sits a large chin-mounted air intake, feeding air directly into the single engine. Its location was carefully designed to avoid turbulent boundary-layer air and reduce the risk of foreign object damage during low-altitude operations. The F-16's powerplant is another key to its renowned performance. Early models were equipped with Pratt & Whitney's F100-PW-200 engine, capable of generating approximately 23,830 pounds of thrust. Later variants of the F-16 featured an upgraded F100-PW-220/229 (for enhanced reliability and thrust) or the General Electric F110-GE-100/129. The engine itself is housed in a removable module, facilitating quick maintenance under field conditions.

The F-16 is a true multirole platform, capable of air-to-air, ground attack, and electronic warfare (EW) missions. This versatility is reflected throughout the Viper's integrated weapons suite, consisting of eleven hardpoints:

six underwing pylons, two wingtip launch rails, two under-fuselage stations, and one centerline station. The wingtip rails typically accommodate air-to-air missiles such as the AIM-9 Sidewinder or the AIM-120 AMRAAM. These rails minimize drag and maximize launch parameters. For ground attack missions, the F-16 also carries a wide variety of precision-guided munitions (PGMs), including the GBU-12; JDAMs (Joint Direct Attack Munitions); and the AGM-65. For close-quarters combat, an M61A1 Vulcan 20mm rotary cannon, firing up to 6,000 rounds per minute, is integrated into the plane's left-side wing root.

But perhaps the most revolutionary aspect of the F-16's structural design is its intentional aerodynamic instability. Most traditional aircraft are designed to be inherently stable in flight. This stability facilitates ease of control, but it comes at the expense of maneuverability. Ironically, the F-16 was engineered to be slightly *unstable*, making it far more responsive to pilot inputs and capable of rapid directional changes. Managing this instability is the role of the fly-by-wire controls. This system electronically transmits the pilot's inputs to the aircraft's control surfaces via quadruple-redundant computers, replacing the older mechanical linkage systems. It enables a more precise control of the F-16's pitch, yaw, and roll, allowing for tight turns and high-G maneuvers. The system also prevents pilot-induced errors (including over-rotations and stalls) by limiting control inputs within safe parameters. All told, this fly-by-wire system marked the beginning of a new era in fighter aircraft design, enabling airframes to be optimized for performance first, with control made possible via software.

Early variants of the F-16 were fitted with the Westinghouse AN/APG-66 pulse-Doppler radar, capable of detecting and tracking airborne targets while simultaneously providing terrain-follow and ground-mapping modes. It also supported the critical "look-down/shoot-down" capability to intercept low-flying aircraft.

Subsequent iterations of the F-16 featured the upgraded AN/APG-68, offering enhanced range, resolution, and multi-target tracking. It enabled BVR missile engagements and introduced Synthetic Aperture Radar (SAR) mapping for improved ground targeting. Many of the modernized F-16s, particularly those undergoing Service Life Extension Programs (SLEP), now include the AN/APG-83 Scalable Agile Beam Radar (SABR). This Active Electronically-Scanned Array (AESA) radar delivers faster target acquisition, higher-resolution mapping, EW resistance, and simultaneous air-to-air and ground-attack modes. These radar evolutions have kept the F-16 relevant in modern threat environments, enhancing both its lethality and survivability.

In all, the F-16's structural design was a revolution in aerospace engineering—a high-performance platform made possible by risk-taking, innovation, and a deep understanding of air combat requirements. From its blended fuselage and agile wing profile to its fly-by-wire controls and ergonomic cockpit, every component was tailored for dogfighting supremacy and multirole flexibility. Though John Boyd himself lamented the increasing weight and complexity added during the initial production, the core of his vision survived. And his vision has endured throughout every iteration of the F-16.

The true testament to the Viper's legacy, however, is not its longevity...but its continued relevance. As upgraded variants fly alongside 5th-generation fighters, the F-16 remains a potent symbol of how thoughtful design and iterative improvements can keep a 20th-Century warplane relevant well into the 21st Century. Central to these iterative improvements is the "Block" system—a way by which the US Air Force and aerospace manufacturers have tracked major upgrades to the airframe, onboard systems, and capabilities. Each Block introduces significant changes: Often including new avionics, engine upgrades, weapons integrations, or revised mission roles.

Beginning with *Block 1*, the initial production variant of the F-16A/B rolled off the assembly line in 1978. It was a lightweight, day-only fighter designed primarily for air-to-air combat. *Block 1* introduced the airframe, fly-by-wire controls, and the APG-66 radar. However, it had limited capabilities for all-weather operations or ground-attack roles. *Block 5* quickly followed, incorporating minor updates such as radome and paint-scheme refinements along with enhanced corrosion protection. Then came *Block 10*, which standardized many of the lessons learned during the early years of the F-16's service. These included an improved wiring system and upgraded cockpit instrumentation. These early F-16s were still relatively simple compared to what the aircraft would become, but they proved to be highly effective and popular among frontline squadrons.

With the onset of *Block 15*, the Viper improved dramatically. Introduced in the 1980s, it featured the larger horizontal stabilizers, improving maneuverability and control at high angles of attack. *Block 15* later integrated the AIM-7 Sparrow missiles for the F-16 ADF (Air Defense Fighter) configuration, thereby increasing the jet's BVR capability. Other *Block 15* ADF upgrades included enhanced radar capabilities and "Identification Friend or Foe" (IFF) systems.

Block 25 introduced the F-16C/D variant in 1984, ushering in a new era of digital avionics, night-fighting capabilities, and advanced weapon suites. This block was delivered with the APG-68 radar, which had a greater range and resolution. Along with improved cockpit displays and an upgraded mission computer, the *Block 25* aircraft could now take on a broader range of missions.

Block 30/32, introduced in the late 1980s, offered an engine choice: General Electric's F110 for *Block 30*, and the Pratt & Whitney F100 for *Block 32*. Although *Block 30/32* wasn't a dedicated SEAD (Suppression of Enemy Air Defenses) variant, it often assumed that role until the *Block 50/52* came online. Early radar and avionics upgrades were also introduced for *Block 30/32*, laying the groundwork for later multirole versions.

Block 40/42 marked the arrival of the LANTIRN (Low Altitude Navigation and Targeting Infrared for Night) system, making the F-16 an all-weather, night-capable strike fighter. These blocks introduced a strengthened airframe, wide-angle HUD, and upgraded landing gear to accommodate heavier payloads. *Block 42* retained the Pratt & Whitney engine, while *Block 40* used the General Electric variant. Together, these blocks formed the backbone of the US Air Force's operations during the Gulf War and beyond.

Block 50/52, introduced in the early 1990s, brought even more enhanced capabilities. These jets were optimized for SEAD missions with systems like the HTS (AGM-88 Targeting System) pod. They featured the APG-68(V)5/9 radar, improved EW systems, and new datalinks. Like the previous *Block 40/42*, this iteration featured different engines for the two stablemates: *Block 50* used the General Electric F110-GE-129 engine, while *Block 52* used the Pratt & Whitney F100-PW-229. Both versions, however, included GPS-guided weapons integration, a more powerful mission computer, and modernized cockpit displays. These aircraft remain in widespread service today and are the standard for many foreign customers.

By the early 2000s, General Dynamics had sold its military aircraft interests to Lockheed Martin. Under Lockheed's new direction, *Block 60* was developed exclusively for the United Arab Emirates (UAE). Sometimes called the "Desert Falcon," this version features conformal fuel tanks, the AN/APG-80 AESA radar, an advanced mission computer, and an internal EW suite. It also includes an upgraded cockpit with color MFDs, and FLIR capability built into the nose. These enhancements arguably made the Desert Falcon the most advanced foreign-sales variant F-16 of its time.

Finally, *Block 70/72* represents the latest and most sophisticated evolution of the Fighting Falcon. Introduced during the late 2010s, and entering service in the 2020s, this version combines the best of all previous blocks with cutting-edge technology. *Block 70* uses the General Electric engine, while *Block 72* uses the Pratt & Whitney variant. Both versions nevertheless feature the AN/APG-83 SABR AESA radar; a modernized cockpit; an automatic ground collision avoidance system; and an enhanced EW suite. These jets also benefit from a SLEP system that increases their structural lifespan up to 12,000 flight hours, thus ensuring decades of vitality and durability.

The evolution of the F-16's "Block" program is a story of smart modernization. While the airframe has remained largely the same, the systems within the Viper have been radically updated to meet changing threats and mission demands. This iterative flexibility has allowed the Viper to stay combat-relevant for more than 45 years—a feat few fighter aircraft can match.

Naturally, the United States was the first and largest operator of the F-16 Fighting Falcon. Today, the US maintains a substantial fleet of Vipers within the US Air Force and Air National Guard. But even before its full introduction into American service, the F-16 attracted international attention. NATO allies, particularly those in Europe, were eager to replace their aging fleets of F-104s. This effectively launched the F-16 into the global spotlight. Over the next few decades, it would become the fighter of choice for more than two dozen air forces worldwide. Its versatility, ease of maintenance, and superb performance made it a favorite among aerial arms from Israel to South Korea.

In 1975, Belgium, the Netherlands, Denmark, and Norway all selected the F-16 under a joint production agreement, marking the first major foreign sale for the aircraft. In the decades that followed, these nations benefitted from the F-16's multirole capability, offering both air-to-air and ground-attack versatility. Belgium's fleet, for example, has repeatedly deployed in support of NATO missions, showcasing the adaptability and enduring relevance of the F-16.

Denmark and Norway, however, recently began phasing out their legacy fleets. Norway, one of the five original NATO partners to receive the F-16, announced that they would replace the Viper with the F-35 Lightning II. Norway initially ordered seventy-four F-16s (*Block 1, 5, 10* and *15*), the first of which arrived at Bodø Airport in

1980. After two decades of various upgrades and conversions, Norway sold thirty-two of their F-16s to Romania in 2023, with the remainder being donated to Ukraine. The Royal Danish Air Force maintained a fleet of more than 70 aircraft, most of which have since been sold to the Argentine Air Force or given to Ukraine.

Further afield, Israel has long been a prime operator of the F-16, with a unique narrative of customization and enhancement. These exclusive updates resulted in the F-16I *Sufa* variant, featuring advanced avionics and an extended operational range.

In Asia, Japan developed its own licensed copy of the F-16 (dubbed the "F-2") built by Mitsubishi Heavy Industries. South Korea, meanwhile, ordered thirty-six F-16C/D *Block 32* aircraft in 1981, making it the first foreign operator of the C/D-model Viper. Over the ensuing years, many of these ROK F-16s were built in-country. Today, the ROK Air Force operates nearly 150 F-16s, but the South Korean government plans to phase out the aircraft in favor of the forthcoming indigenously-developed KF-21 fighter. The Singaporean Air Force's acquisition of the F-16 in 1988 signified a significant upgrade in Southeast Asia's aerial defense capabilities. Enhanced by locally supported upgrades and training programs, Singaporean F-16s reflect the city-state's commitment to maintaining a strategically viable, high-tech air force. Likewise, Taiwan's procurement and operation of the F-16 since 1992 underscores the strategic balance of power in the Pacific Rim.

In South America, Chile and Venezuela illustrate the F-16's reach and adaptability across different continents. Chile, through the *Peace Puma* program, incorporated the Viper into its air force, enhancing its own national defense in the post-Pinochet era. Meanwhile, Venezuela's acquisition of the F-16 reflected its strategic priorities, despite more-recent geopolitical complications that have impacted the plane's operational status.

The F-16's operational journey also extends to the Middle East, where Egypt and the UAE continue to capitalize on the plane's multirole functionality. Egypt's fleet (consisting of multiple Blocks) has regularly participated in peacekeeping efforts while supporting national defense, most recently targeting ISIS camps in Libya in 2015. Similarly, the UAE's investment in the newer *Block 60* "Desert Falcon" demonstrates their ongoing commitment to regional security and air power projection.

For Greece and Turkey, the F-16 plays a significant part in balancing regional tensions within the Eastern Mediterranean. Both countries received the F-16 in the late 1980s, and have since undertaken extensive modernization programs to enhance their fleets. For the Turkish Air Force, F-16s have been the undisputed workhorses of homeland air defense, intercepting their Greek counterparts over the Aegean Sea and targeting insurgencies along the outskirts of Asia Minor.

For many of the world's air forces, the F-16 has been more than just a means of homeland air defense—it's been a symbol of international cooperation and technological evolution. Its widespread implementation underscores the mutual defense strategies and technological adaptability that have made it a mainstay of modern air power. Indeed, the F-16's legacy is one of resilience and transformation, proving that lightweight fighters can, in fact, *redefine* air superiority. Lessons learned from its operational history continue to influence current and future endeavors in military aviation.

Today, the F-16 Fighting Falcon remains in production, having flown in conflicts from the Middle East to the Balkans and beyond. But its story begins in the backrooms of the 1970s—a time of reckoning, reformation, and rebirth in American air power. It is the story of how a ragtag group of rebels changed the trajectory of aviation history, and built the Falcon to fly for generations to come.

F-16I "Sufa" from the Israeli Air Force's 253rd Squadron at Ramon Airbase, 2016. *Israeli Air Force*

Chapter 2:
Wings over the Negev

From its first combat missions in 1981 to its critical role in the 2021 Israel-Palestine Crisis, the F-16 has long been at the forefront of Israeli air power. Emerging victorious from the Yom Kippur War in 1973, the Israeli Air Force (IAF) knew that victory alone wasn't enough; they had to evolve or risk being outflanked. The Mirage III and F-4 Phantom had carried them through some of the deadliest dogfights in modern history, but the clock was ticking on their dominance. Meanwhile, bruised but unbroken, the Arab states were treating their wounds and turning to Moscow for a new breed of MiGs—faster, deadlier, and more advanced. The next war would be fought with 4th-Generation fighters, and the Israelis weren't going to wait for the first shot to be fired. Hence, the IAF sought delivery of the F-15 Eagle and F-16 Fighting Falcon. With the F-16 and its heavier F-15 stablemate, the IAF didn't just maintain air superiority; they *redefined* it.

When the IAF first took delivery of the F-16 in July 1980, few could have predicted how quickly the aircraft would earn its place in history. Within months, the jet—nicknamed "Netz" (Hebrew for "hawk")—drew its first blood in the skies over Lebanon and Iraq. Indeed, by mid-1981, Israeli F-16s had destroyed the Osirak nuclear reactor in Baghdad and scored their first aerial victories against the Syrian Arab Air Force (SyAAF). These events were not just tactical victories; they were geopolitical milestones that signaled a new age of air combat in the Middle East.

First Engagements: April 1981

The early 1980s were a turbulent time in the Levant. The Republic of Lebanon, once praised as the "Switzerland of the Middle East" had collapsed into a bloody civil war. As central authority disintegrated, the country became a battleground for competing factions. The Palestine Liberation Organization (PLO), Syrian occupation forces, and a patchwork of Lebanese militias all vied for control amid the chaos. For Israel, the stakes were growing deadlier by the day. PLO militants had entrenched themselves in southern Lebanon, using the region as a launchpad for rocket attacks into northern Galilee. At the same time, Syria's military presence in Lebanon raised alarms in the halls of Tel Aviv and Jerusalem. To Israeli policymakers and military planners, the combination of border attacks and Syrian entrenchment represented not just a security threat, but a broader regional challenge that demanded a decisive response.

The IAF, meanwhile, remained on high alert. In fact, by the time Lebanon's civil war exploded in 1975, Israeli pilots had been flying routine air patrols over Lebanese territory, where dodging anti-aircraft fire and sampling enemy MiGs had become a regular occurrence. It was against this backdrop that the IAF gradually integrated the F-16 into its operational tempo, running high-speed intercepts and ground attack missions with increasing regularity. Though the F-16's first sustained combat mission came during Operation *Opera*—the infamous strike against the Osirak facility on June 7, 1981—the Netz would achieve its first air-to-air victories three months earlier.

On April 28, 1981, the IAF launched a routine aerial surveillance patrol near the town of Zahle in the Bekaa Valley. Zahle had recently been the site of a Syrian military buildup, and the IAF was determined to monitor Syrian intentions. During this mission, Israeli ground control intercept (GCI) radar detected two slow-moving, low-altitude targets flying from Syrian-controlled territory into the central Bekaa Valley.

But these radar contacts were unusual; they didn't match the speed or altitude profiles of a fixed-wing aircraft. Based on their flight characteristics (and the most-recent intelligence data), the slow-movers were identified as Mi-8 helicopters. These twin-turbine, medium-lift choppers were often used for resupply missions or delivering ground troops to the battlefield.

GCI scrambled the nearest F-16s onto the contacts. Answering the calls were Major Dubi Yoffe and Lieutenant Rafi Berkovich, both from the 117th Squadron, flying separate missions aboard the F-16s designated *Netz 126* and *Netz 112*, respectively. Within their respective formations, Yoffe and Berkovich were operating under strict rules of engagement: Visual confirmation was necessary before firing.

Vectoring to intercept, Yoffe and his wingman approached the first helicopter from a higher altitude, descending through the broken clouds to acquire visual contact. This first Mi-8 was flying relatively low (possibly below 1,000 feet), maneuvering lazily over the rugged terrain. Whether it was carrying troops, cargo, or conducting a low-level reconnaissance mission remains unclear.

What *is* clear, however, is that the Mi-8 was a potential

threat.

And, under the rules of engagement, Israeli pilots were authorized to engage any Syrian aircraft within Lebanese airspace. Yoffe's F-16, armed with a full contingent of AIM-9 Sidewinders, closed within a few miles before locking on to the target.

Seconds later, Yoffe's missile found its mark.

The hapless Mi-8 erupted into flames, breaking apart mid-air while scattering debris across a nearby hillside.

Meanwhile, on the other side of the Bekaa Valley, Rafi Berkovich (the IAF's youngest pilot at the time), steadily locked on to the second Mi-8. Barely ten miles from the Riak Airfield, he had gained radar contact of an unidentified, slow-moving aircraft. As he accelerated to gain visual confirmation, however, Berkovich suddenly lost the contact on radar. Undeterred, he continued his flight pattern until he reacquired the target, confirming that it was indeed an Mi-8.

Locking on to the target, Berkovich calmly fired his AIM-9. "The missile left the plane," he recalled, "on the left side with a whoosh, and I followed it with my glance." His anticipation quickly turned to disgust, however, when he saw the AIM-9 vector downward, hitting the ground and blowing up a nearby shack. Later that day, Israeli news media reported that the IAF had launched rockets into the area, mistakenly believing that Berkovich's errant missile had been a rocket attack.

Frustrated but determined not to let the Syrian chopper get away, Berkovich drew closer, thumbing his selector switch from "Missiles" to "Guns." Pulling into the six o'clock position, he squeezed off a multi-round burst, which sent the helicopter careening into the desert floor, trailed by ominous plumes of black smoke.

The downed helicopters of April 28, 1981 were historic kills for the IAF: The first air-to-air victories for Israeli F-16s, and the first such victories for *any* F-16 worldwide. However, the true test of the Viper's power would come later that summer, during one of the most audacious air raids in military history.

Operation *Opera* (1981)

On the afternoon of June 7, 1981, a flight of Israeli F-16s streaked low across the desert skies, thundering towards Baghdad. Their mission: To destroy the Osirak nuclear reactor before it could produce weapons-grade plutonium. Codenamed Operation *Opera*, this bold airstrike was a masterclass in strategic foresight, military airmanship, and technological audacity. It was also the first time Israeli F-16s appeared in a major, sustained combat operation.

But *Opera* was more than just a tactical airstrike—it was a bold, preemptive attack with enormous geopolitical implications. At its heart was the deeply-held belief that nuclear weapons should be kept out of the hands of hostile regimes.

By the late 1970s, Saddam Hussein had emerged as one of Israel's deadliest adversaries. Under his dictatorial leadership, Iraq's government had been openly hostile towards Israel, rendering aid and financial support to the PLO.

But the bigger concern was Iraq's nuclear ambition.
France had recently given Iraq an *Osiris*-class research reactor, yielding the nickname "Osirak" (a portmanteau of "Osiris" and "Iraq"). On paper, the Osirak facility was intended for domestic nuclear research. But the Israeli intelligence community (particularly Mossad) wasn't

convinced. They suspected Iraq's endgame was to extract plutonium for a nuclear bomb.

Adding to this sense of urgency was the regional instability following the Iranian Revolution of 1979. Saddam, sensing a regional power vacuum and fearful that the Ayatollah Khomeini's rhetoric would galvanize Iraq's Shiite majority, invaded Iran in September 1980. Despite the ensuing Iran-Iraq War, Israeli analysts feared that Saddam was pursuing a nuclear weapon not to deter Iran, but to challenge Israel.

As early as 1980, Prime Minister Menachem Begin convened Israel's top military and intelligence leaders to formulate a plan. His driving principle would come to be known as the Begin Doctrine. In other words, Israel would not allow any hostile regime to acquire nuclear weapons, no matter the diplomatic or military cost.

Covert operations like sabotage and targeted assassinations had already failed. In April 1980, for example, Mossad agents in France allegedly helped sabotage reactor components before their delivery. In June of that year, an Egyptian scientist working on the reactor was assassinated in Paris. But these efforts only delayed the inevitable. By 1981, Israeli intelligence concluded that the reactor would soon become "hot," thereby making any future airstrike too dangerous due to the potential for radioactive fallout.

An immediate strike, therefore, seemed to be the only option.

For the operation, IAF leaders devised a plan that was both complex and audacious. It would be a long-range, low-level airstrike penetrating deep into Iraqi territory. The operation would involve eight F-16 Netz fighters, each loaded with two 2,000-pound Mk-84 unguided bombs, escorted by six F-15 "Baz" fighters for top cover. The strike package would need to fly more than 600 miles from southern Israel, across Saudi and Jordanian airspace, to strike Saddam's reactor in broad daylight before returning. To avoid radar detection, the F-16s would have to fly at dangerously low altitudes, down to and including 100 feet.

Mission planning occurred on a need-to-know basis and under terms of absolute secrecy. Even within the IAF, only a small group of pilots knew the exact nature of the target. To the outside world, the operation had been disguised as a routine, long-range training mission. The pilots involved were ordered not to tell their families (or fellow officers) about the true nature of the mission.

Leading the strike package would be Lieutenant Colonel Zeev Raz, a decorated F-4 Phantom pilot who had recently transitioned to the F-16. Raz was selected not only for his flying skills but for his judgment and composure under fire. His wingmen were some of the most elite pilots in the IAF, including: Amos Yadlin, Ilan Ramon, Amir Nachumi, and Relik Shafir—names that would become legendary within the Israeli defense community.

And every pilot knew the stakes.

"If this goes wrong," said Raz, "we're not coming back. But if it works, it will save the country from a nuclear nightmare."

At around 4:00 PM local time on June 7, 1981, the strike package launched from Etzion Air Base in southern Israel. The F-16s, flying in a tight formation, descended to tree-top level almost immediately to avoid the Saudi and Jordanian radar envelopes. Their accompanying F-15s, meanwhile, flew at higher altitudes, acting as decoys, escorts, and early warning sentinels.

Under strict radio silence, the F-16 pilots navigated their way using pre-programmed INS (Inertial Navigation Systems) and dead reckoning. Below them lay the vast emptiness of the Saudi and Jordanian deserts—hostile airspace where even a moment's detection on radar could trigger a high-speed intercept. Ironically, the day's biggest threat came not from enemy radar, but from a vacationing monarch. As the planes crossed into Jordanian airspace, King Hussein bin Talal was on holiday in the coastal town of Aqaba. He reportedly spotted the planes overhead and recognized the prominent roundels which bore the Star of David. Unsure of whether the Israelis were intending to attack him, the Saudis, or Baathist Iraq, he urgently radioed his Air Command. But the king's warnings were either delayed or ignored. For by the time Iraq's radar had detected the incoming jets, it was already too late.

At approximately 5:35 PM, the F-16 strike package emerged from the desert, ascending rapidly to a medium altitude for their bombing run. The Osirak facility lay just ahead—a prominent dome surrounded by scaffolding and other pieces of infrastructure, part of which were still under construction.

Zeev Raz was the first to dive.
Rolling his F-16 into a 35-degree dive, he aimed his aircraft directly at the reactor dome, releasing both of his Mk-84 bombs from an altitude of 6,000 feet. Behind him, the other seven F-16s followed in close succession, each pilot timing their delivery within seconds of one another.

Their Mk-84s were configured with delayed fuses so the bombs could penetrate the reactor's superstructure before detonating. The first wave of impacting bombs tore through the outer building, while the second wave collapsed the reactor's containment structure, resulting in a massive fireball that erupted from the center of the dome. In less than two minutes, the IAF pilots had accomplished their mission. But destroying the reactor was only the first step. They needed to get out of Iraqi airspace alive.

The Israelis quickly vectored south, accelerating to full afterburner. A few Iraqi SAMs and air defense guns reflexively opened fire, but none of the weapons found their mark. Minutes later, Iraqi fighters scrambled to intercept the Israeli bandits, but by then, the F-16s were already at the edge of Saudi airspace. All fourteen Israeli aircraft returned safely, landing at around 7:00 PM to a stunned and jubilant reception.

The pilots who flew in Operation *Opera* later reflected on the geopolitical impact of their mission. Amos Yadlin, who later became the head of Israel's Military Intelligence Directorate, recalled the moment he saw the reactor explode beneath him: "There was no doubt in my mind, we had done something that would echo for decades. We were striking not for glory, but for survival." Ilan Ramon, the youngest pilot on the mission and Israel's first astronaut (who tragically perished in the Space Shuttle *Columbia* disaster of 2003), said in a later interview: "We flew in total silence. I remember looking down and seeing the Tigris River shimmering in the sun. I was just 27, and I understood the magnitude of what we were doing. We weren't just flying a mission. We were flying into the heart of a threat to our nation's existence. We all knew what was at stake. When we pulled up and saw the reactor dome vanish in a ball of fire, it was a surreal moment."

The strike had been an operational success: the Osirak reactor was destroyed, and Saddam's nuclear ambitions were set back by at least a decade. Reactions from the international community, however, were sharply divided. The UN Security Council, for instance, condemned the attack, declaring it a violation of international law. Nevertheless, Operation *Opera* remains a pivotal moment in the history of the Levant. It was the first preemptive airstrike to target a nuclear facility, setting a precedent followed by the United States and its allies for decades to come. Tactically speaking, *Opera* demonstrated the F-16's capability as a precision strike aircraft, especially during long-range missions. Operationally, it showcased the IAF's disciplined approach to training, planning, and execution. Strategically, it reinforced Israel's doctrine of preemption and deterrence. And it sent a clear message to adversaries across the Middle East that Israel would act decisively to neutralize any perceived threat. But most of all, Operation *Opera* underscored a simple truth: A handful of Israeli pilots, flying American-made jets, had reshaped the balance of power in the Middle East.

July 1981: F-16 vs. MiG-21

Barely one month later, the F-16 would prove its mettle yet again, this time in a classic fighter duel. Although still riding high from the success of Operation *Opera*, Israel's contention with Syria continued unabated throughout the summer of 1981. Following the loss of their Mi-8 helicopters in April, Syrian forces deployed more of their SAM batteries into the Bekaa Valley. The Israelis, meanwhile, increased their operational tempo, flying more missions to gather intelligence and maintain air superiority.

On July 14, 1981, IAF radar stations detected flights of Syrian MiG-21s vectoring towards Lebanese airspace, presumably to challenge Israeli air patrols. With growing concern over Syria's intentions, the IAF had been sending more Combat Air Patrols (CAPs) of F-15s and F-16s to assert control over the airspace and deter any SyAAF incursions.

Major Amir Nachumi (who had been a subordinate flight leader during the raid on the Osirak reactor) went aloft in his F-16 that morning as part of the daily CAP cycle. According to Nachumi's own account, the mission began as a standard CAP station, patrolling the contested airspace near the Lebanese-Syrian border. While on patrol, GCI radar picked up a flight of Syrian MiG-21s approaching from the northeast.

Then, the order came quickly: Intercept and Engage.
"All of a sudden, we got a vector to the northeast," Nachumi later recalled in an interview. "They told us MiGs were airborne. We were ready. We had trained for this."

As the F-16s vectored in, radar operators at the command center relayed real-time updates. Nachumi spotted the enemy aircraft with his own eyes as they thundered towards the Israeli formation. "It was a clear day. I looked and I saw a dot—just a dot at first—but it grew larger and faster than anything else in the sky," he said. "I knew right away it was a MiG."

The MiG-21, whose pilot remains unknown, tried to engage from a higher altitude, diving into the fight while gaining momentum. But Nachumi, relying on the F-16's agility and superior turn rate, maneuvered into an offensive position.

"It was like a dance," Nachumi continued. "You can't

lose sight of your opponent for even one second. He tried to shake me, but every time he turned, I turned tighter."

Accelerating to nearly Mach 1, Nachumi closed the distance. "I got a tone from my Sidewinder," he recalled, readying the AIM-9 mounted on his wing. "The seeker was locked. I squeezed the trigger and watched the missile leap off the rail."

The Sidewinder tracked perfectly, slamming into the rear fuselage of the MiG-21. The explosion was visible for several miles, and the Syrian fighter disintegrated in mid-air. Nachumi never saw a parachute.

"It was over in seconds," he recalled. "A short dogfight. But in that moment, all the training, all the missions, everything came together. It was clean, decisive. That's what we train for."

Downing the Syrian MiG marked another significant milestone in the F-16's combat history. Nachumi had become the first F-16 pilot to shoot down an enemy fixed-wing fighter in combat. In many ways, it confirmed the Viper's superiority over 3rd-Generation Soviet-built fighters and gave IAF commanders a renewed confidence in the Netz. But Nachumi's kill was also symbolic, proving that the IAF could counter Syrian air threats without incurring catastrophic losses.

Amir Nachumi, already a decorated veteran from the Yom Kippur War, later referred to this kill as one of the cleanest and most precise of his career. "There's no joy in killing another pilot," he admitted. "But there is satisfaction in knowing you did your job. You protected your country. That's what matters."

This dogfight from July 1981 was one of many skirmishes in the prelude to the larger aerial campaigns of June 1982, when the IAF would achieve total air superiority over Lebanon. It also cemented Nachumi's reputation as one of the IAF's most skilled and battle-tested fighter pilots.

Beyond its tactical success, however, this encounter demonstrated the shifting technological balance in the Middle East. The MiG-21, once considered a potent adversary in the skies over Vietnam and the Six-Day War, had now become obsolete in the face of Western 4th-Generation fighters like the F-16. As Nachumi later said during a retrospective interview: "The MiG was still dangerous. You had to respect it. But the F-16 gave us the edge. It let *us* dictate the fight, not the other way around."

Spring Dogfights: April-May 1982

By the spring of 1982, the skies over Lebanon had become the focal point in the ongoing tensions between Syria and Israel. Syrian air defense troops were reinforcing their SAM networks, while the SyAAF had dispatched their MiG-23s—the "swing-wing" interceptors capable of high-speed BVR engagements. Israel, for their part, responded by launching deeper reconnaissance missions and more-aggressive fighter patrols to suppress Syrian threats.

On April 21, 1982, an Israeli E-2 AWACS detected a flight of Syrian MiG-23s heading towards the Bekaa Valley. The 117th Squadron responded, scrambling a number of F-16s to intercept. Among them were Lieutenant Colonel Zeev Raz (the flight leader from Operation *Opera*) and Captain Hagai Katz, respectively flying the F-16s tail-numbered "107" and "284." *Netz 107*, as Raz's plane was called, had previously been piloted by Amos Yadlin during the attack on the Osirak reactor the year before. Under various pilots, *Netz 107* would go on to achieve six and a half enemy kills, a world record for the most air-to-air victories attributed to a single F-16.

As the Vipers approached the Lebanese border, AWACS vectored Raz and Katz to intercept the incoming MiGs. The Syrian pilots, hoping to catch the F-16s off-guard, flew in at lower altitudes to avoid radar detection.

But it was no use.
The F-16s' look-down/shoot-down capabilities had already seized the advantage. In the resulting high-speed melee, Zeev Raz maneuvered *Netz 107* into a clear line of sight, locking onto the lead bandit with an AIM-9 Sidewinder, and skillfully downing the MiG-23 with a clean, one-shot kill. Moments later, Katz locked onto the second MiG. Employing the same tactics, Hagai Katz silenced the remaining MiG with a hot Sidewinder.

Tensions escalated further on May 25, when another dogfight erupted between Israeli Vipers and Syrian MiG-21s. That afternoon, IAF radar detected Syrian MiGs operating west of Damascus. From their flight characteristics, the bogeys appeared to be on a CAP station or reconnaissance patrol over eastern Lebanon, part of their broader efforts to monitor Israeli activity along the border. In response, the IAF scrambled a flight of F-16s to intercept. Among the responding pilots that day was Lieutenant Colonel Amos Mohar from the 110th Squadron.

Flying lead in his two-ship formation, Mohar and his wingman vectored toward the radar contact. According to post-mission reports and subsequent interviews, Mohar got a tally on two MiG-21s that were already performing evasive maneuvers. Indeed, the Syrian pilots realized they were being hunted, and were now attempting to disengage or perhaps maneuver themselves into a better position.

Mohar would later describe the engagement as "brief but very intense." In a 2003 interview, he recalled: "I saw the first MiG pull hard to the right. I went into afterburner and followed. He was trying to force an overshoot, but the F-16 is tight in the turn. I managed to keep him inside my HUD and fired a Sidewinder at close range."

The first kill came rather quickly.

Mohar's AIM-9 tracked perfectly, impacting the MiG's rear fuselage, which sent it spiraling into the ground. The second MiG-21 attempted to disengage, diving low in a panic to avoid visual detection and in hopes of confusing the F-16's radar.

Mohar, undeterred, stayed in pursuit.

"The second MiG tried to get low," he continued, "but I had the energy advantage. I got behind him, and at about 4,000 feet, I locked him up and fired another Sidewinder. This one also connected." The second MiG exploded mid-air, confirming Mohar's second kill in less than two minutes. With this engagement, Mohar had earned the rare distinction of a "double kill" in a single sortie—a feat that less than a handful of IAF pilots had achieved up to that point. Syria, meanwhile, confirmed the loss of two MiGs but gave no official word regarding the pilots' fates.

Mohar later referred to this mission as one of his most intense aerial engagements. "There's no time to think," he said. "It's all instinct and training. And on that day, it all worked as it was meant to." But most of all, the engagement had further solidified the F-16's reputation as a formidable dogfighter in the IAF.

As it turned out, this incident of May 25, 1982 would be the final prelude to a larger conflict that erupted just a few weeks later. As part of their ongoing efforts to eradicate the Syrian-PLO alliance, Israeli forces launched Operation *Peace for Galilee* on June 5, 1982. During that conflict, Israeli F-16s would go on to dominate Syrian MiGs in a series of historic dogfights over the Lebanese countryside.

Operation *Peace for Galilee* (1982)

Operation *Peace for Galilee* marked a turbulent phase in Lebanon's Civil War. Although the PLO was still Israel's primary target, *Peace for Galilee* soon escalated into a much broader confrontation against Syrian forces. In fact, *Peace for Galilee* grew to such a size and intensity that, today, many historians consider it to be a war unto itself—often cited as the "1982 Lebanon War."

But whatever name the conflict may take, it was nevertheless a major combined-arms offensive that rivaled the scale of the Yom Kippur War. The stated objective was to drive PLO forces north of the Litani River, thus eliminating their base of operation for rocket attacks into northern Israel.

By the second day of the conflict, Israeli ground forces had penetrated deep into southern Lebanon. Meanwhile, Syria sent aloft a sizeable fleet of MiG-21s and MiG-23s to contest Israeli air operations. The IAF, in turn, deployed their F-15s and F-16s to run fighter patrols over the Bekaa Valley, where Syrian bandits (flying from bases inside Syria and Lebanon) had become a growing nuisance.

First Blood: June 8, 1982

On June 8, the stage was set for the first major dogfights of the air campaign. That day, Israeli F-16s were tasked with patrolling Lebanese airspace near the town of Zahle, where Syrian fighters were known to operate.

At the tip of the spear was the 110th Squadron, with Major Dubi Ofer and his wingman, Captain Avishai Canaan, flying a two-ship patrol near the southern end of the Bekaa Valley. Ofer was flying lead when a pair of MiG-23s appeared on his radar. The MiGs were likely attempting to intercept Israeli reconnaissance aircraft or provide close air support to Syrian ground troops. According to IAF records, Dubi Ofer engaged the lead MiG, successfully scoring an AIM-9 kill.

Avishai Canaan, meanwhile, turned onto the second MiG-23.

Carefully maneuvering to gain a rear-aspect position on the Syrian MiG, Canaan fired his own AIM-9 Sidewinder. The missile glided effortlessly into the tail of the enemy bandit, sending the MiG into a flat spin before crashing into the valley floor below. But these shootdowns were just the first in a broader series of engagements on June 8 wherein the IAF destroyed multiple Syrian aircraft with no losses of their own.

Later that day, Major Moshe Rosenfeld, a pilot with the 253rd Squadron, skillfully downed a MiG-21 over the Bekaa Valley. "The sky was full of aircraft," he recalled. "The MiG tried to break right and dive away, but I had tone and launched. The missile hit squarely. No chute, no wreckage I could follow—just a puff of smoke and then nothing." This description, provided for an IAF Oral History Project, illustrates the lethal precision of the AIM-9 Sidewinder missile when launched under optimal parameters.

Shortly thereafter, Rosenfeld engaged a second MiG-21, this time alongside his wingman, Captain Arnon Sharabi. According to after-action reports and IAF mission logs, both pilots fired at the fleeting MiG simultaneously.

Sharabi remembered: "Rosenfeld called out the bandit. I had tally as it broke low and left. I locked and fired just seconds after he did. The MiG was destroyed, but we both saw the hits. It was impossible to say whose missile struck first." As such, the IAF credited *both* men with a shared kill, officially marking it as a half-victory for each pilot—a rare but not unprecedented occurrence in the history of Israeli air power.

Elsewhere over the Bekaa, Lieutenant Shlomo Sas, flying his F-16 with the 110th Squadron, drew tally on another Syrian MiG-23. According to IAF debriefs, Sas engaged this MiG-23 while it was attempting to position itself for a missile shot on another Israeli aircraft. But, leveraging the superior performance of his F-16 and its AIM-9 Sidewinders, Sas quickly outmaneuvered the MiG, got a solid tone, and fired. The missile struck the MiG-23's fuselage dead on, sending the Syrian bandit down in flames.

These early victories set the tone for the remainder of the air campaign. Meanwhile, the F-16's performance underscored why it had become *the* weapon of choice for aerial engagements. Over the next few days, they would achieve more than forty air-to-air victories against Syrian MiGs, effectively neutralizing the threat and dominating the skies over Lebanon. Beyond air-to-air combat, however, the F-16s easily switched roles, performing ground-attack missions whenever needed. This versatility was crucial, as the Israeli Vipers targeted Syrian SAM batteries and disrupted the enemy's attempts to re-fortify their positions.

Gaining Momentum: June 9, 1982

On the morning of June 9, the skies were clear over Lebanon, offering high visibility as IAF fighters crossed into the Bekaa Valley. Over the course of this single day, more than a dozen aerial engagements unfolded, each one a high-speed duel of tactics and split-second decisions. Among the pilots flying that day was Captain Eliezer Shkedi of the 117th Squadron, seated in the cockpit of *Netz 107*. At approximately 12:00 PM, in the northern Bekaa Valley, Shkedi's radar registered an incoming contact.

A pair of MiG-23s had been vectored onto his patrol. He identified their approach at 20,000 feet, closing in fast from the west. Rolling his F-16 into a high-G maneuver, Shkedi descended to leverage the sun at his back, blinding the enemy pilots momentarily. This textbook "energy maneuvering" allowed him to maintain his airspeed while closing in range.

As the trailing MiG-23 tightened into a sharp left turn, Shkedi followed suit, demonstrating the F-16's tight turning performance. As he later recalled: "The G-forces pushed me deep in my seat. Every move mattered. I flew instinctively, but my mind raced with checklists. Check the radar, check the weapons, keep visual."

Lining up his sight for a missile shot, Shkedi pulled the trigger, launching his Sidewinder from a range of nearly 1,500 meters. The missile tracked straight into the target, striking the MiG-23 directly in its engine.

Debris scattered as the fleeting MiG burst into flames. Within less than sixty seconds from his initial contact, Eliezer Shkedi had scored his first kill of the day—a testament to the intense training and cool-headed airmanship of the IAF fighter jockeys. Throughout the engagement, their comparative altitudes fluctuated between 15,000 and 18,000 feet, a band that favored the F-16's climb rate and the MiG-23's tendency to lose energy quickly under tight turns. In fact, by the 1980s, it was a well-known fact that the MiG-23 was a capable interceptor, but a terrible dogfighter. As one US Marine pilot remarked: "The MiG-23 was a maneuvering dog. Even the F-4 Phantom could out-turn it."

Two minutes later, Shkedi and his wingman, Lieutenant Eitan Stibbe, picked up radar contact from another MiG-23. Likely, this was the second MiG from Shkedi's earlier engagement, who was now circling back for an attempted ambush. Stibbe and Shkedi made visual contact on the Syrian bandit at around 23,000 feet. While Shkedi lured the MiG with a feint to the north, Stibbe closed in from below and behind. Describing their teamwork dynamics, Stibbe recalled: "It's about seconds and trust. When Shkedi rolled, I went with him. We almost didn't need words. We had to act as one mind. Every second, we updated each other on position, fuel, and bandit movements."

Determined not to let this MiG get away, the two F-16s began executing "bracket" maneuvers, splitting their formation on either side of the MiG to constrict its evasion tactics. Shkedi pulled a high-G turn to force the MiG into a climb, while Stibbe anticipated the escape trajectory. Together, they sandwiched the MiG-23, forcing it to bleed energy while exposing its vulnerable rear aspect. Both pilots fired their AIM-9 Sidewinders simultaneously; but only one of the missiles struck, sending the MiG-23 into a spiraling descent.

Neither pilot was sure whose missile had downed the bandit. Thus, the IAF happily split credit for the kill, with each man given one-half of the victory. Stibbe later recalled that their tight teamwork and intense training had carried them through the engagement. "The teamwork felt like choreography under fire," he said. "Pressure was off the charts, but we never lost sight of our goal."

The news of Shkedi's and Stibbe's victory skyrocketed morale within the neighboring 110th Squadron as they went aloft over the Bekaa Valley. As the day unfolded, a new wave of MiG-21s entered the fight, for which the 110th Squadron's F-16s were armed and ready. On this latest combat patrol, Major Amir Nachumi (the victorious pilot from the July 1981 dogfight), was flying lead when he encountered the first wave of MiG-21s at 16,000 feet.

Nachumi noticed three MiG-21s vectoring aggressively towards his twelve o'clock position. Keying the radio, he called out:

"Contact! Three bandits. Stay tight."

With his radar in "Search" mode, he maneuvered his F-16 to leverage the sun, tilting left to minimize his signature. At this altitude, the air density gave his F-16 an accelerative edge, and he quickly closed the distance. The MiG pilots, meanwhile, broke formation—one jinking hard right, another descending, while the third began climbing. Nachumi focused on the descending MiG, anticipating an attempt to draw him low and slow. Hesitating for only an instant, he recalibrated his radar, switching from "Search" to "Track While Scan."

"He's not pulling me down with him," said Nachumi.

Instead, he traded altitude for speed, rolling into the turn and engaging his afterburner. The Syrian pilot responded with a tight turn, but the F-16 held a tighter arc at this speed, courtesy of its fly-by-wire system. Nachumi launched an AIM-9 Sidewinder from 1,200 meters. "You aim, you see the lock, and you listen for the growl." The missile struck behind the MiG's cockpit. A brief fireball erupted, followed by an oblong parachute deploying against the cloudless blue sky.

Minutes later, in another sector of the 110th Squadron's patrol zone, Captain Avi Lavi confronted his own MiG-21 not in a chase, but in a near head-on scenario at 18,500 feet. Lavi, on the right side of his formation, picked up radar returns showing a MiG-21 maneuvering to bracket him. Recognizing the threat, Lavi veered off the formation line, dropping to 17,000 feet to disrupt the MiG's targeting solution.

Along with his wingman, Lavi flew into a defensive split, drawing the Syrians into a two-on-two engagement. The MiG fired a missile shot, but Lavi's hard left break and chaff deployment caused the weapon to miss its mark. Leveling off, he caught a glimpse of the Syrian pilot pulling into a high "yo-yo"—a defensive maneuver designed to slow an attacker's approach while conserving one's own energy. Realizing that the MiG was trying to position itself into a better attack angle, Lavi timed his reversal perfectly, pulling inside the MiG's turn, and placing himself squarely into the firing window.

"Everything slows down. You hear your heartbeat. You see your target, and suddenly, it's just you and him." Switching to missile lock, Lavi waited for a clean tone from his Sidewinder. The missile cut through the air and detonated under the MiG's right-side wing root. Lavi's voice, steady and elated, echoed over the radio net:

"Splash one!"

From contact to kill, the engagement lasted less than two minutes, further highlighting the F-16's agility in a turning fight.

Meanwhile, farther north, Lieutenant Roee Tamir occupied the critical number four spot in his flight's diamond formation. His relative position allowed for a clear view of the battlespace, a vital advantage at 15,500 feet, as GCI reported another round of Syrian launches from the southwest. His radar soon registered a pair of MiG-21s, attempting to sneak below the main engagement. Tamir maintained close communication with his lead, relaying continuous altitude and bearing calls. Observing the MiGs trying to slip underneath his radar envelope, he descended his F-16 with controlled urgency. Glancing at his weapons panel, Tamir selected an AIM-9, using his onboard radar to cue the seeker.

He closed within 1,000 meters before uncaging the missile. Seconds later, the Sidewinder met its mark, sending the MiG hurtling into the valley below.

Elsewhere over the Bekaa, Captain Relik Shafir (another veteran of Operation *Opera*) sat alert in his F-16, acting as the flight leader for another 110th Squadron fighter patrol. Suddenly, Shafir's radar sounded the alert of an incoming contact.

The next few minutes quickly devolved into chaos. Dozens of radar contacts appeared as Shafir's formation sallied forward at 500 knots. As Shafir recounted later: "There's no luxury for confusion. Eyes and mind must split in several directions...all at once." While sweeping across the valley's western sector, he caught a blip on his radar: two Syrian MiG-21s ascending rapidly to intercept from the northeast.

The pair of MiGs soon streaked into view, their slim profiles vectoring into a head-on angle. Closing at high speed, Shafir and the MiGs climbed to 23,000 feet.

But the F-16 got its energy advantage first.
Shafir quickly designated the leading MiG as his primary threat. With his radar locked and the distance narrowing to less than five miles, he readied his AIM-9 Sidewinder. Steadily yanking up the nose of his F-16, Shafir brought his seeker onto the hot target aspect of the MiG-21.

The MiGs broke formation, one peeling up and away, the other continuing a shallow climb. Shafir throttled down momentarily, adjusting his closure rate, then committed as the MiG crossed into his HUD. The AIM-9 growled a solid tone. "I remember thinking, you have seconds to choose. Missile or guns, high angle or chase. You trust your training; it all narrows to the target." He squeezed the trigger at just under two nautical miles. Flames lanced from underneath his wingtip, and the screaming Sidewinder tracked right into the rear of the MiG, sending it down under a spiraling trail of smoke.

With one bandit down, Shafir began searching for the other MiG.

But the engagement zone was already teeming with missile trails and exploding targets. He dove hard, evading an incoming missile from another direction, firing his flares, and jinking right to spoil the missile's lock. Maintaining his spatial awareness was the result of instinct and disciplined training, as he was constantly checking his six while listening for directives. "Every beep, every flash on the radar, it's a threat or an opportunity," he said. "You become obsessed with what's behind you but never forget what's ahead."

A few hours later, during a follow-on mission, Shafir caught another radar spike: a Syrian MiG-21 on a defensive orbit, attempting to maneuver into his blind spot at 18,000 feet. The enemy pilot dove aggressively, hoping to force an overshoot, while exploiting the MiG-21's smaller turning radius at lower speeds. But Shafir used the F-16's advanced flight controls to counter, pulling into a high-G "scissors" maneuver, while carefully maintaining his energy state. His airspeed slowed to 350 knots as he forced a rolling scissors, luring the MiG into a vertical extension.

With the F-16's radar maintaining a soft lock, and a quick switch to the AIM-9, Shafir out-rolled the MiG, regaining his positional advantage. At approximately one mile separation and dead astern, he obtained missile lock and fired. The Sidewinder connected flawlessly, with the MiG-21 trailing fragments of its airframe as the pilot ejected.

Both encounters highlighted critical facets of air-to-air combat: discipline, technological supremacy, and a relentless focus on threat prioritization. Shafir's two kills in a single large-scale operation reflected not only his individual airmanship, but the critical synergy between man and machine. "There's a moment you know it's not just you up there; it's your squadron, your training, your aircraft...all woven together," he later said. Managing multiple simultaneous threats, Shafir was able to distinguish genuine dangers from fleeting radar ghosts. Each decision (including when to break and when to fire) was grounded in a rigorous threat assessment and the calm, collected confidence of an expert.

Later that day, Major Ofer Einav (also from the 110th Squadron) was leading a four-ship flight, armed with a full contingent of AIM-9 Sidewinder missiles. His initial warning came from a nearby AWACS: "Bandits inbound, angels twelve, bearing one-seven-zero."

As Einav's F-16 sallied forward at 18,000 feet, his onboard radar—set to "Track-While-Scan" mode—picked up multiple returns. Suddenly, Einav spied a visual contact: two MiG-21s whose airframes glistened in the sunlight below, crossing eastbound at 12,000 feet, and separating from a larger flight formation.

Climbing to bracket the threat and maintain their energy advantage, Einav led his own formation into a high-speed dive, maintaining his airspeed at nearly 520 knots. At 6,000 feet separation, Einav locked on to the leading MiG-21, confirming missile tone. He later recalled: "I saw the glint of the afterburner and tried to predict his break."

Selecting his AIM-9, Einav visually tracked the MiG as the Syrian pilot jinked hard left. Rolling out, he fired his missile from a range of 1,850 meters, maintaining a shallow climb to keep the F-16's energy up and its sensors locked. Moments later, the Sidewinder tracked flawlessly, impacting the MiG-21's rear fuselage. The kill confirmation came seconds later:

"Splash one!"

The MiG began billowing smoke, then spiraled beneath a broken cloud layer until Einav visually saw the pilot eject.

Meanwhile, Captain Avishai Canaan (who had downed a MiG-23 the day before), was three miles behind Einav's section when his threat receiver went off, indicating a MiG-21 at his ten o'clock. Flying at 15,000 feet, he engaged his radar in "Range-While-Search" mode. "It was a matter of seconds. You learn to trust your radar and your instincts," Canaan recounted during his official IAF interview. He visually acquired the MiG-21 as it pulled a hard G, climbing through a trailing cloud. Canaan assessed the MiG to be about 3,500 meters off, head-on at more than 1,000 knots closure.

He selected his AIM-9, and switching to "Single Target Track," Canaan received a solid tone and hit the trigger. "Fox 2!" he cried, as the missile launched through the air.

Instantly, the MiG-21 began a defensive left break, dropping flares, but it was too late. The missile detonated its proximity-fuse mere feet from the enemy bandit. Seconds later, flames erupted from the MiG, which spiraled downward and quickly disappeared from view. Visual confirmation from Canaan's wingman, along with AWACS radar loss of the hostile contact, confirmed the kill.

The final kills over the Bekaa Valley that day belonged to Lieutenant Colonel Uri Gil and Lieutenant Nimrod Goor, both from the 253rd Squadron. Uri Gil was flying his F-16, *Netz 290*, as the wingman in a two-ship formation. Shortly after entering Syrian airspace, his radar flashed the signature of two Syrian MiG-23s heading southward. Gil throttled *Netz 290* into a split-S maneuver, descending rapidly to lower the radar cross-section and disrupt the enemy's ability to track him. He gained visual contact at approximately 14,000 feet, noting the glint of the MiGs' canopies in the sunlight.

Gil's first engagement began as the MiGs attempted to bracket the Israeli pair. Arming his Sidewinder, Gil pushed the plane into a hard left turn, forcing the Syrian pilot to commit. Pulling 7.5 Gs, he came in behind the lead MiG-23, gaining a position of tactical advantage. "In the heat of battle, seconds felt like minutes. My focus narrowed to the tone in my headset and the shape of the bandit ahead," Gil later recalled. He fired the Sidewinder from nearly a mile out, tracking the contrail as the AIM-9 made its first kill of the day.

The remaining MiG went into survival mode, firing its afterburner and breaking low. Gil followed the fleeting bandit through a series of aggressive turns, matching the MiG's evasive maneuvers with calculated bursts of throttle and flap adjustments to sustain the F-16's energy. Within moments, Gil artfully downed the second MiG-23 in similar fashion. The Syrian pilot ejected, his canopy arcing skyward as the aircraft spun out below.

Reflecting on both encounters, Uri Gil noted the superior situational awareness offered by the F-16's avionics suite. "Nothing in the simulator prepares you for the real target in your sights…the smoke, noise, and a heartbeat pounding in your ears," he told his debriefers. He also emphasized the constant need to update target angles, energy states, anticipated missile threats, and cited the F-16's agility as the critical asset that allowed him to neutralize two adversaries in short order.[1]

Flying lead in a separate formation, Lieutenant Nimrod Goor reported an equally-intense dogfight against another MiG-23. At just below 20,000 feet and 480 knots, Goor locked on to a radar contact designated by the local AWACS. Minutes after Uri Gil's victories, Goor's situational read of the enemy's tactics allowed him to anticipate the MiG climbing high, attempting to break out

---

[1] Gil was no stranger to the horrors of combat. By 1982, he was one of Israel's most seasoned pilots, having downed an enemy MiG-21 during the Six Days War. He retired from active service in 1985, but remained a Reservist until 2003, retiring for good at the rank of brigadier general.

above the Israeli fighter sweep.

Goor thus took his F-16 into a high-G vertical reposition, trading speed for altitude to intercept the MiG's flight path. "I could see the MiG-23 turning hard, trying for altitude advantage. I had to counter every move, or risk losing him," Goor later shared. At the merge, Goor rolled out below and behind the MiG, outmaneuvering the aircraft as it attempted to reorient. Choosing to fire the AIM-9 at minimum range, he watched as the missile tracked and detonated just aft of the cockpit, resulting in a clear kill.

His debrief highlighted the quick adaptation to changing angles and the need for strong visual identification, since radar contact had become unreliable among the shifting ground clutter. As Goor noted: "Every engagement is unpredictable," thus emphasizing the need for disciplined teamwork during moments of chaos. The fog of war (often intensified by conflicting radio chatter and missile alerts), forced him to rely on his instincts, training, and spatial awareness to survive.

In their post-mission analysis, the 253rd Squadron recognized the combined effects of strong leadership, effective tactics, and the ability to exploit fleeting targets. All of the above had decisively contributed to achieving air superiority. Gil's and Goor's performance that day not only reflected their individual skill and courage, it underscored broader lessons about the teamwork, discipline, and technical proficiency that were necessary to dominate modern aerial engagements.

Each engagement from June 9, 1982 highlighted the pilots' individual mastery of the F-16 and its weapons, along with the importance of teamwork, communication, and decisive action. These early successes carried their momentum into the following day's engagements, whereupon Israeli F-16s would tally up another handful of aerial victories.

Day of Aces: June 10, 1982

On the morning of June 10, 1982, now-Captain Rafi Berkovich (the young IAF pilot who downed one of the Syrian Mi-8s in April 1981) was flying back into the Bekaa Valley, his radar painting blips that moved with dangerous intent. At 9:34 AM, Berkovich's first engagement of the day started when his radar picked up a pair of inbound Syrian fighters. Identifying the bandits as MiG-23s, Berkovich squared his F-16 into a climb, gaining the altitude advantage. The Syrian pilots were flying low, using the ground to mask their approach, but Berkovich chose a steep angle of attack, climbing past 20,000 feet to minimize exposure to ground fire and maximize his radar perspective.

Berkovich then received permission to engage.
As the pair of MiG-23s turned, they split defensively. Berkovich targeted the leader, firing off a Sidewinder, and watched as the missile glided effortlessly into the MiG's tailpipe. With one MiG destroyed, he narrowed his focus onto the second.

The remaining bandit jinked hard right, but Berkovich rolled his F-16 to match the MiG's evasive actions. "He tried to shake me," Berkovich recalled, "but the F-16 turns tighter every time." Four seconds later, he fired off another Sidewinder, scoring a direct hit. The Syrian fighter exploded into a ghastly orange fireball before descending into the valley below.

Fresh on the heels of this double-kill, at just after 10:05

AM, a MiG-21 appeared fast and low, coming from the northeast. As the MiG-21s were lighter and leaner than the MiG-23s (built for quick dogfights and short reaction time), Berkovich immediately shifted tactics, lowering his altitude and leveraging the F-16's tighter turning radius.

The MiG-21 executed a scissors maneuver, trying to force Berkovich out in front for a missile shot. He countered by tightening his circle, using the advantage in the F-16's thrust-to-weight ratio. Thumbing his selector switch to "Guns," Berkovich fired off a quick burst of 20mm rounds, striking the MiG dead astern, before sending its peppered airframe down through the clouds below.

Meanwhile, Berkovich's squadron mate, Captain Sasha Levin, was hunting MiGs in a different patrol zone. The morning light glinted off his F-16's canopy, and the first radio calls of the day alerted him to multiple enemy vectors. At 10:22 AM, his radar tagged a low-flying bogey: a MiG-21 darting in and out of the ground clutter. Levin closed to visual range, using intermittent afterburner to mask his approach and blend in with the background heat signatures. The Syrian pilot dove hard, trying to escape by flying close to the rough terrain. Levin stayed close, matching every move, but stayed at a higher altitude, waiting for the bandit to commit to a turn. As expected, the MiG pulled up, allowing Levin to get a firm lock-on.

The resulting AIM-9 shot impacted the MiG just as it was leveling out of the climb.

Levin's next encounter was even more unusual—a SA-342 Gazelle helicopter, armed and flying low to the ground north of the valley. Attack helicopters posed a unique threat: Slow-moving, often hidden by the terrain, but armed with deadly anti-tank missiles. Tracking the SA-342 with his onboard radar, Levin took the F-16 into a rolling dive to close the distance. He then saw the helicopter pop up nearly 300 feet above the valley ridgeline. He maneuvered from behind, before firing another AIM-9 that shredded the Gazelle's tail rotor.

By now, the airspace over the Bekaa was saturated with air-to-air kills. MiGs, Gazelles, and ground fire had filled nearly every sector of the valley. The intensity grew with each passing moment, as more pilots from the 117th and 110th Squadrons entered the fray. Weather conditions remained favorable for visual engagements. Although broken clouds had cast shadows over the hilly terrain, visibility above 15,000 feet remained virtually limitless.

Captain Shlomo Zayteman (117th Squadron), as a section leader, was vectoring northeast at 22,000 feet when he received a GCI call:

"Target, thirty miles, bearing zero-five-zero. MiG-23."
Zayteman, maintaining a level profile, switched his APG-66 radar to "Range-While-Search" mode. Confirmation came moments later as the MiG-23's radar return appeared on his scope.

At approximately 9:12 AM, Zayteman banked left, reducing throttle just enough to keep visual on the horizon as he trimmed his F-16's nose for optimum angle-of-attack during ingress. The MiG-23, flying a conservative sweep at 18,000 feet, attempted a gentle descending turn. Zayteman then dropped flaps, increasing lift as he transitioned into a high-speed pursuit at over 450 knots. With the MiG-23's afterburner visible, Zayteman armed his AIM-9 Sidewinder at a six-o'clock low approach, remaining just out of the MiG's rear-quarter view to avoid detection. At more than one mile separation, the

Sidewinder acquired tone.

Zayteman pulled the trigger.

The missile tracked clean and the Syrian MiG exploded in a brilliant fireball, debris falling in a scattered arc north of Rashaya. Zayteman vectored away from the wreckage, scanning for additional threats, as his element regrouped. He later recalled in a 1983 interview: "All that training took over in those seconds."

By this time, Captain Ami Lustig, also from the 117th Squadron, was already maneuvering southwest at 19,000 feet. The weather was slightly more turbulent at this altitude, with thin bands of cirrus clouds affecting forward visibility. Lustig monitored his threat receiver, cautiously aware of the ongoing missile launches detected from Syrian SAM sites. At 9:21 AM, GCI tagged a fast-moving MiG-23 crossing south of Lake Qaraoun, altitude 15,000 feet and descending. Lustig accelerated, deploying chaff in anticipation of ground-based missile threats, and toggled his F-16's radar to "Air Combat Maneuvering" (ACM) mode. He then gained visual contact on the bandit just beyond a patch of sun glare.

Careful not to over-commit, Lustig rolled into a lead-pursuit attack geometry at about 500 knots. The MiG-23 attempted a high-G break to the right, to which Lustig responded by pulling vertical, sustaining 7.5 G's before rolling inverted and regaining his firing position at 14,200 feet. He fired a quick multi-round burst from the autocannon, striking the MiG-23's rear fuselage. At 9:22 AM, Lustig confirmed his kill over the net and continued a wide sweep, checking for further engagements.

Meanwhile, Hagai Katz, on patrol with a different flight

element, flew at 23,500 feet near the southern edge of Baalbek. Katz paid close attention to the shifting tactical picture, frequently scanning both visually and on radar. Describing his approach in a 1999 interview, Katz described a landscape that was "littered with hostile radar locks and friendlies crossing at all altitudes." At 9:31 that morning, he spied a MiG-23 at nearly 11,000 feet, 12 nautical miles east and climbing. Katz vectored to intercept, using the sun's position to mask his angle of approach.

Closing at more than 950 knots, Katz armed his AIM-9 and steadied the F-16 to within 1,300 meters. The MiG pilot, aware too late, panicked and pulled hard right. But Katz's higher energy state allowed him to outmaneuver and slip into a tracking position directly behind the MiG. Taking the shot, Katz released his Sidewinder just as the MiG initiated a second break.

"Fox 2!" Katz called.

The missile guided true, striking the MiG-23 mid-air over the ridgeline east of Hermel. Katz reported, "negative parachute," before rejoining his formation and egressing west.

However, no other pilot on June 10 came close to matching Amir Nachumi's kill record for the day. Indeed, as the flight leader for an afternoon mission, he was credited with *five* air-to-air kills in a single day, earning him the rare and coveted distinction of "Ace in a Day." On this latest mission, flying *Netz 237*, Nachumi's element breached Syrian-controlled airspace at approximately 2:02 PM. His objective: Support the suppression of Syrian SAM batteries while destroying airborne threats.

Shortly after entering hostile airspace, his APG-66

radar painted two fast-moving MiG-21s approaching head-on at high altitude, closing from the northwest. At approximately 28,000 feet, with his airspeed steady near Mach 0.95, he locked up the nearest MiG at nine nautical miles, and visually identified the bandits at six miles out. The enemy pilots executed a shallow descent, likely attempting to increase their speed for a missile engagement. As Nachumi recalled in his debrief: "They were bold, coming straight at us. I knew timing was everything." He coordinated with his wingman, maneuvered into a hard right bank, and fired off an AIM-9.

The missile struck the leading MiG-21 at just under two nautical miles. Nachumi saw the engine erupt into flames; the pilot ejected moments later. His voice, captured in mission tapes, was calm: "One MiG down, searching for second." The trailing MiG-21 broke hard left, diving low to escape. Nachumi pursued, engaging the target at a lag pursuit angle to bring the Sidewinder's seeker steadily onto the MiG. He depressed the trigger, and the missile tracked left before connecting. He confirmed his kill at 18,000 feet, over the southern edge of the valley.

During a follow-on mission later that day (and now flying *Netz 234*), Nachumi received an AWACS call identifying a formation of MiG-23s north of Lake Qaraoun. With his F-16 closing fast at 500 knots, Nachumi once again selected his AIM-9 Sidewinder.

From their respective aspects, he realized that the MiG-23s were trying to force him into a turning duel.

But his F-16 had the advantage.

Nachumi accelerated, pulled for altitude, then rolled vertically, pinning his quarry at 22,000 feet from a high six o'clock position. At less than two miles, he fired his

Sidewinder, which tracked right into the first MiG's tailpipe, marking his third kill of the day. The explosion left a black streak as the pilot attempted to eject. "The MiG-23s were aggressive," he later recalled, "but they couldn't match our vertical performance."

Events compressed quickly as Nachumi's radar indicated another contact. This MiG-23 launched its defensive flares, cutting across Nachumi's nose and forcing him briefly into a rolling scissors maneuver. "With multiple bandits, you train yourself not to fixate." He employed a high-G barrel roll, maintaining airspeed through coordinated rudder and throttle inputs, then lined up his final shot. He fired his fourth Sidewinder from a range of 1.5 miles, and his visual confirmation followed: the MiG-23 spun out and crashed northeast of Bar Elias.

With four kills under his belt for the day, a fifth encounter unfolded in the chaotic aftermath. Nachumi's element had trapped a MiG-23 between their positions. His wingman fired, but Nachumi's coordinated maneuver forced the MiG into an unrecoverable split-S at low altitude. The enemy aircraft crashed before either Nachumi or his wingman could fire another missile. As Nachumi reported: "Sometimes the best weapon is to box them in, let them make a mistake."

All five air-to-air kills (including his contribution to the crash of the fifth aircraft) were confirmed by the IAF and credited to Nachumi. His precise employment of the AIM-9, mastery of the F-16's agility, and composure under fire showcased the apex of Israeli air combat skills. Nachumi later told debriefers: "The F-16 gave us confidence, especially against the MiG-23. Our systems, training, and teamwork made the difference when seconds counted."

Final Kills: June 11, 1982

Throughout the following day—June 11, 1982—the dance of missiles and maneuvers continued unabated, slicing the heavy air with trails of heatseeking AIM-9s and bursts from the rattling 20mm cannon. Although the conflict was far from over, June 11 would be the final day of the F-16's aerial engagements for the remainder of the 1982 Lebanon War. Each pilot's perspective from this final day of dogfighting provides a story of precision, courage, and the harsh realities of air combat in one of the most challenging operational theaters of the era.

That morning, Lieutenant Roee Tamir (110th Squadron) piloted his F-16 at 27,000 feet, scanning through the bright, cloudless skies over Lebanon. Intelligence briefings foretold an increased SyAAF presence, and Tamir's formation was flying a fighter sweep to protect Israeli ground forces from aerial attack. At approximately 9:51 AM, his radar picked up a fast-moving contact at fourteen miles out, approaching head-on and descending rapidly.

Tamir's adversary, a MiG-21, attempted to close the gap for a missile shot. Tamir, recalling his textbook dogfighting tactics, dropped to 25,000 feet and increased his airspeed to 380 knots. He broke hard right at the merge, forcing the MiG pilot into a high-G turn. With this maneuver, however, the MiG began losing energy and altitude, allowing Tamir to lock on with his AIM-9 Sidewinder. He later recalled: "I saw the MiG's afterburner flare, a desperate attempt to break off. I squeezed the trigger at 1.2 miles." The missile tracked unerringly and struck the MiG's tail section, sending the airframe into an uncontrollable spin as black smoke billowed from its mechanical innards. Tamir maintained visual, confirming the kill before rejoining his wingmen, his voice steady as he reported: "Splash one MiG-21."

Meanwhile, Eitan Stibbe was airborne once again, flying over central Lebanon when his threat receiver sounded the dreadful tone of an enemy radar lock. He immediately performed a split-S, diving through scattered clouds at 400 knots, hoping to break the enemy's lock. As it turned out, Stibbe had caught a signal from a MiG-23 approaching from 2 o'clock low at 16,000 feet. The MiG-23 could accelerate beyond Mach 1 in level flight, but the F-16's agility allowed Stibbe to stay out of sight until the last moment.

At six miles out, Stibbe banked left and then abruptly reversed, creating separation. The MiG-23 attempted a head-on pass and fired its R-23 missile, but the shot was wide. Stibbe then rolled inverted and pitched downward, gaining position behind his opponent's tail at 3,000 meters. Selecting his AIM-9, he recalled: "It was about reading the MiG's intentions. The moment he pulled up, I was already closing and got tone. I fired at 2,000 meters, closed to confirm the kill." The missile struck just behind the cockpit, exploding the MiG into a violent fireball.

Elsewhere over Lebanon, Stibbe's young teammate, Lieutenant Rani Falk, flew in a tight three-ship formation at 18,000 feet. Midday turbulence bounced the trio of F-16s as the squadron neared Rayak Airfield, where hostile contacts had been tracked. Suddenly, Falk spotted a MiG-21 climbing a fast leftward arc at 12,000 feet, maneuvering for an intercept vector.

Falk initiated a descending left-hand spiral, closing at a 45-degree angle while keeping the MiG in sight. "He tried a vertical scissor. I matched with maximum G and stayed outside his turn radius." The dogfight tightened as Falk fired a short burst from his M61 Vulcan at 700 meters, hitting the MiG's right wing. "My tracers walked right across his wingtip. I saw the panel fly off."

The MiG wobbled, trailing fuel.
And Falk watched as the pilot ejected, the aircraft plummeting eastward.

Later that afternoon, Major Dani Oshrat (117th Squadron) entered combat at 20,000 feet, his radar registering multiple bogies. Engaged by a MiG-21 at medium range, Oshrat performed a barrel roll to force the MiG into an overshoot, then quickly rolled out to get behind his adversary. Oshrat fired off a Sidewinder from nearly two miles out, impacting the MiG-21 halfway into its climb. Seconds later, a Sukhoi Su-22 fighter-bomber passed below, heading toward Israeli armored positions. Oshrat rapidly descended, closing into range. A second AIM-9 Sidewinder from his wingtip cratered the bandit's fuselage, sending the Sukhoi down in flames.

At around the same time, Lieutenant Amos Bar was vectored by ground control toward another pair of Su-22s entering his sector at low altitude, moving along the valley floor to evade radar. Bar descended aggressively, the F-16's afterburner roaring as he tracked his target visually. Closing fast to within 1,000 meters, he thumbed his selector switch to "Guns," lining up the reticle of his M61 autocannon. Bar kept steady at less than 500 meters from the Su-22 as he squeezed the trigger. The F-16 shuddered from the burst as his tracer rounds pierced the sky. "I focused on keeping the reticle fixed," he recalled, "knowing I had just seconds to get it right."

As Bar recounted in his mission debrief: "Everything was happening at treetop level. The Su-22 tried to jink left, but I stayed on his tail. I fired, saw him break left, hit the deck, and crash in a cloud of dust." Even as the explosion echoed and debris scattered, Israeli pilots began to realize that air superiority was nearly within grasp—setting the stage for the deepening battle unfolding over the Bekaa Valley.

As twilight settled over the valley, Rani Falk was back in action, aloft in his F-16 on another fighter sweep. It was just after 5:00 PM, and radar intercepts had picked up a new mass of Syrian fighters entering the valley at low altitude. Falk's radio crackled with an urgent transmission: A flight of Su-22s was skimming the treetops south of Baalbek.

Falk dropped to 2,000 feet and keyed his radar, acquiring the nearest Sukhoi at 2.5 nautical miles. With the sun behind his wings, Falk maneuvered for a tail chase, maximizing his own visibility and minimizing the Su-22's chances of spotting him. Selecting his infrared missile, he waited until the Su-22 was in a hard right turn at about 330 knots. "I got a perfect aspect, closed to just within a mile," Falk later recounted. "The tone screamed in my headset. I squeezed off the missile and saw it hit just aft of the cockpit." The Syrian pilot ejected, his aircraft plowing into the fields below, and Falk immediately rejoined his formation, eyes scanning for further threats.

The combat tempo accelerated as Eitan Stibbe, flying his second sortie of the day, turned southwest along the ridgeline northeast of Zahle. At 4:49 PM, GCI vectored Stibbe towards a pair of low-flying Su-22s at approximately 4,000 feet. As he leveled at Mach 0.9,

Stibbe locked onto the lead aircraft at a range of 1.2 nautical miles. He fired his Sidewinder, watching the missile arc under the Syrian's left wing and detonate against the instep of the engine. Flames engulfed the Su-22, which spiraled downward under a heavy plume of black smoke. "There was no time to celebrate," he recalled, "a second Su-22 was turning inside my arc." Stibbe pulled into a high-G barrel roll, rolling in behind the second Syrian fighter. Closing within 1,000 meters, he switched to guns and fired a two-second burst, scoring critical hits across the left wing. The Syrian pilot ejected, tumbling through the afternoon air as his aircraft disintegrated beneath him.

His radar humming with alerts, Stibbe pressed southward and spotted a distinctive silhouette: an SA-342 Gazelle helicopter. Flying low over the orchard-lined flats at less than 200 feet, the Gazelle presented a small, elusive target. Stibbe, hunting at treetop level, sampled the target with short radar bursts to avoid giving away his position. At an airspeed of 350 knots, he closed to within 900 meters and triggered his AIM-9, hitting the helicopter's tail boom and sending it crashing into a grove of cypress trees.

During that same hour, Relik Shafir vectored towards a flight of Syrian Su-22s skirting along the western escarpments of the Bekaa Valley. At 4:58 PM, Shafir and his wingman intercepted the Sukhois in trail formation. The Syrians dove low and fast, trying to shake their pursuers among the wadis, but Shafir stay hot on their tail. As one Su-22 tried to pull away at full afterburner, Shafir halted the bandit's escape with a deft AIM-9 missile shot. The pilot ejected seconds before the ailing jet cartwheeled on impact with the ground.

Shafir immediately targeted the remaining bandit who, now alerted, attempted a hard evasive maneuver.

Shafir, however, anticipated the turn.

Banking hard left, he fired his second Sidewinder. The missile tracked flawlessly and exploded beneath the right wing, sending the wayward Sukhoi nose-first into the hillside.

Flying to the north, Lieutenant Yehuda Bavli was covering the valley's upper reaches. With scattered clouds at 3,000 feet and excellent visibility under evening conditions, Bavli tracked a solitary Su-22 crossing his nose at about 700 knots. Timing his vertical maneuver, he pulled up and rolled inverted, dropping behind the target. Bavli locked onto the bandit at a distance just beyond one nautical mile, launching the missile straight for the Su-22's heat signature, striking amidships. "You could see the flare from the hit through the cloud base," Bavli recalled. The Su-22 crashed into the open countryside, and Bavli returned to base with minimal fuel remaining.

The detailed accounts of these Israeli F-16 pilots' air-to-air victories reveal not only their skill and courage, but also the evolving tactics that shaped aerial combat during the Bekaa Valley campaign. Each engagement highlighted how cutting-edge technology and rigorous training can shape the outcome of aerial combat. These aerial victories, in turn, influenced the evolution of Israeli air combat doctrine and pilot training in the years that followed.

Although Operation *Peace for Galilee* highlighted the efficacy of Israeli air power, the conflict itself ended somewhat inconclusively. By late June, Israeli forces, together with their Christian Lebanese allies, had occupied the capital of Beirut. Under pressure from Israeli bombardment, the PLO eventually negotiated a ceasefire. PLO forces were subsequently evacuated from Lebanon under supervision from a multinational peacekeeping force. After removing the Syrian-PLO influence, Israel installed a friendly, Christian-led government in Lebanon headed by President Bachir Gemayel.

Sadly, Gemayel was assassinated in September 1982, just weeks after taking office. With Gemayel's demise, the chance of signing a permanent peace treaty became increasingly unlikely, and Israeli public opinion began to sour on the war. Thus, the IDF withdrew from Beirut and formally ended its occupation on September 29, 1982—the date upon which most historians agree as marking the end of the 1982 Lebanon War.

The resulting armistice of May 17, 1983 provided for an orderly withdrawal of Israeli ground forces. Over the next two years, IDF troops gradually withdrew to their South Lebanon "security zone" along the Israeli border, consolidating their final positions on June 5, 1985. The Israeli troops would remain in the security zone until their final withdrawal from Lebanon in 2000.

Aftershocks: June 1985

The skies over Israel remained relatively quiet for the rest of the 1980s. For the next twenty-four years, there would be only *one* F-16 air-to-air engagement until the start of the Second Lebanon War in 2006. That engagement—occurring on June 13, 1985—saw Israeli F-16 pilot Lieutenant Colonel Yitzhak Gat achieve an aerial victory over a hostile reconnaissance drone. This unmanned aerial drone was among the first of its kind—identified as a Tu-143 (DR-3), believed to have been operated by Syrian forces or an aligned militant group such as Hamas or

Hezbollah. Although the 1982 War was ostensibly over, local Arab forces occasionally prodded the area, sending surveillance assets into and around southern Lebanon. This engagement of June 13 marked one of the earliest known instances of a manned fighter jet destroying an unmanned aerial vehicle in combat with an air-to-air missile.

At the time of interception, Gat was flying a routine patrol over the Bekaa Valley as part of Israel's continued effort to monitor enemy activity. The drone was detected flying at a medium altitude, conducting what appeared to be a surveillance sweep. GCI radar relayed the contact and vectored Gat's F-16 to intercept.

Gat made visual contact, confirming the bogey as a DR-3, and launched an AIM-9 Sidewinder after hearing the shrill tone of his missile lock. As expected, the AIM-9 made short work of the enemy drone, destroying the interloper in mid-air.

During later interviews, Gat remarked on the unusual nature of the engagement. He recalled that it was a strange target—not something a pilot would expect to encounter when flying an F-16. The IAF knew that Syrian-PLO forces were using drones, but most Israeli pilots had only seen them in pictures. Still, these early-model drones were a threat, and Gat treated it as such.

The incident underscored Israel's early awareness of UAV threats and its readiness to adapt conventional fighter tactics against emerging aerial systems. It also foreshadowed the rise of UAVs as a critical and cost-effective means of air power. Over the ensuing decades, these aerial drones would play an ever-growing role in conflicts from Bosnia to Afghanistan to Ukraine.

Reflecting on the engagements from 1981-85, these sophisticated dogfights and low-level intercepts demonstrated the exceptional performance and versatility of the F-16: Amassing more than 40 air-to-air victories without a single loss. Israel's strategic use of their F-16s, underscored the aircraft's dominance and effectiveness in both long-range and close-combat scenarios. This was not mere chance, but a deliberate effort characterized by disciplined training and lessons learned from earlier missions. Their success stories echoed beyond the immediate theater, cementing the F-16's reputation as a formidable fighter.

However, the brilliance of Israeli F-16 operations during the 1982 Lebanon War also extended into ground-attack missions, highlighting the F-16's dual capabilities, with pilots often shifting from air-to-air engagements to ground targets in a seamless transition. This dual role underscored the aircraft's adaptability—a critical attribute given the dynamics of modern warfare. Israeli F-16s carried out extensive missions aimed at disabling enemy infrastructure and military assets. These missions included suppressing SAM batteries, identifying and destroying artillery positions, and interdicting supply routes. These ground-attack sorties illustrated not just the impact of the F-16, but its essential part in the wider framework of coordinated military strategy.

Moreover, the operational knowledge and experience gained from the F-16's missions laid the groundwork for subsequent improvement in the IAF's combat tactics, strategy, training and technology. This proactive approach to analyzing and reinforcing operational success ensured Israel's combat readiness for future conflicts. Not surprisingly, when the Second Lebanon War erupted in

2006, it was clear how the lessons learned by Israeli pilots in 1982 had shaped all subsequent fighter operations.

2006 Lebanon War

After Israel completed its final withdrawal from Lebanon in 2000, tensions remained relatively low until the summer of 2006. On July 12 of that year, Hezbollah militants launched a cross-border raid into northern Israel, killing three IDF troops and capturing two others. Israel responded with a massive combined-arms campaign—Operation *Change of Direction*—to eliminate Hezbollah's enclave in Lebanon. The ensuing 34-day conflict came to be known as the "Second Lebanon War," or simply the 2006 Lebanon War.

Almost immediately, the IAF was in action over the Lebanese countryside. From July-August 2006, Israeli F-16s flew more than a thousand combat sorties, most of which were ground-attack missions against Hezbollah's regional infrastructure: Ammunition warehouses, rocket and artillery batteries, Command & Control centers, and bunkers throughout southern Lebanon. For these missions, F-16s made extensive use of the Mk-84 general-purpose bombs and, for higher-value targets, the GBU-12 Paveway II laser-guided bombs.

Still, there was plenty of air-to-air combat against the growing fleet of enemy drones. For example, on the afternoon of August 7, 2006, Israeli air defense radar detected a Hezbollah reconnaissance drone (later confirmed to be an Iranian-built Ababil UAV) approaching from southern Lebanon. Answering the call, an Israeli F-16 occupying a nearby CAP station, vectored to intercept.

As documented in IAF after-action reports, the pilot visually identified the target over the coastal highlands of northern Israel.[2] Aligning his aircraft for optimal firing parameters, he readied himself for a missile shot. Reports vary as to whether the pilot fired an AIM-9 Sidewinder or the domestically-built Rafael Python missile. Both were heat-seeking munitions, and both would have been ideal for engaging UAVs under the current parameters. But whichever missile he ultimately selected, the Israeli pilot destroyed the UAV and prevented it from completing its mission—a fact confirmed in the IAF's operational summaries.

Just before dawn on August 13, 2006, IAF radar detected a second Ababil drone flying low over the Lebanese border. Two F-16s scrambled from Ramat David Airbase to intercept. Only minutes into the air, the flight leader visually identified the drone and fired his Python-5 (the most advanced version of Rafael's heatseeking missiles), killing the remote-controlled interloper with a single shot.

These near back-to-back engagements against militant drones highlighted the new technical and procedural challenges faced by the IAF. The flight profiles for these UAVs included low-altitude ceilings and erratic patterns, designed to evade ground-based radars and limit conventional missile lock capabilities. Intercepting these drones required the F-16 to operate at reduced speeds using precise radar modes and extensive visual scanning. Despite the capabilities of the F-16's missile suite, target acquisition of small drones demanded unusual piloting techniques and a heightened situational awareness. These engagements also exposed the relative vulnerability of

---

[2] Current IAF Operational Security guidelines prohibit disclosing names of pilots involved in counter-terrorism operations.

airspace to even modest UAV intrusions, especially when adversaries employed commercial off-the-shelf and military-grade systems blended within civilian air traffic patterns.

Following the August 13 intercept, the IAF conducted an extensive review of their engagement protocols for countering UAVs. When the war ended, after-action reports discussed the evolving sophistication of Hezbollah's UAV fleet. The F-16s' performance validated Israel's investment in all-aspect air-to-air weaponry, but it also revealed gaps in aerial surveillance and early warning coverage. Operational debriefs recommended integrating enhanced UAV-detection algorithms into airborne and ground radars, which led to the rapid delivery of additional low-altitude detection systems at strategic airbases.

The 2006 Lebanon War, meanwhile, ended almost as inconclusively as its 1982 forbearer. UN Resolution 1701 brokered a ceasefire on August 14, with no clear winner or decisive change in territory. Militarily, Israel resolved to shorten its airborne response times. The war also accelerated the development and deployment of air defense systems like the Iron Dome and David's Sling. In the decades that followed, the IAF maintained higher levels of surveillance over Lebanon and continued to refine its counter-UAV tactics.

Gaza War (2008-2009)

The peace that followed the Second Lebanon War was short-lived. Throughout 2008, the relationship between Israel and Hamas (the new *de facto* governing authority in the Gaza Strip) had taken a turn for the worse. After Hamas's victory in the 2006 Palestinian legislative elections and its subsequent takeover of Gaza in June 2007, the divide between the Fatah-led Palestinian Authority in the West Bank and Hamas in Gaza sharpened regional and international tensions. Israel, citing security concerns and continued rocket attacks on its southern towns, implemented a blockade against Gaza.

During the first half of 2008, both Israel and Hamas engaged in a series of reciprocal attacks. Rockets and mortars from Gaza were met with targeted airstrikes and low-level incursions.

On June 19, 2008, Egypt brokered a truce between Israel and Hamas, which lasted a mere six months before hostilities resumed. Throughout the summer and fall, each side accused the other of violating the terms of the truce. By December 2008, however, both sides had begun escalating military preparations. Israel began massing troops along the Gaza border, moving key air and armored units into position. Inside the Gaza Strip, Hamas mobilized its military wing, the Izz ad-Din al-Qassam Brigades. Local commanders reinforced their air defense capabilities (piecemeal as they were) and intensified efforts to rally public support.

Diplomatic exchanges failed to break the cycle of escalation. Egypt tried to revitalize negotiations, while the UN called for restraint. But neither Hamas nor Israel responded favorably. On Christmas Eve 2008, Israel authorized military action behind closed doors. Finally, on December 27, amidst the spike in rocket attacks from the Gaza Strip, the Israeli government launched Operation *Cast Lead*, beginning with a wave of F-16 airstrikes targeting Hamas' military infrastructure in Gaza.

At 11:30 AM, on the first day of the operation, the IAF flew more than a hundred F-16 sorties. The main units involved were the 107th and 109th Squadrons, respectively based at Hatzerim and Ramat David Airbases, with elements from the 117th Squadron providing supplementary attack and assessment roles.

For their ground-attack missions, F-16 pilots delivered laser-guided GBU-12 Paveway II and GPS-guided GBU-31 Joint Direct Attack Munitions (JDAMs) against pre-identified Hamas targets. A limited number of AGM-65 Maverick air-to-ground missiles were also used for high-value, time-sensitive targets. Standard loadouts for initial sorties included two GBU-31s and two AIM-9 Sidewinders for self-defense, with secondary platforms available for follow-up attacks if necessary. All munitions chosen for the initial wave were selected to minimize collateral damage while maximizing impact against the designated targets.

The first wave launched at 11:32 AM, focusing primarily on Hamas Command & Control centers in Gaza City and Khan Younis. The 107th Squadron, known for its deep-penetration strike missions, led the initial assault. Within three minutes, their F-16s had obliterated a police headquarters, some high-level planning cells, and multiple underground weapons depots. By 11:38 AM, a second wave from the 109th Squadron began targeting the enemy's support infrastructure, including rocket assembly points and artillery caches in Beit Lahiya and Rafah.

Weather conditions remained optimal throughout the entire mission. Clear, cloudless skies and minimal crosswinds allowed for both high-altitude ingress and visually-confirmed target sites. Pilots reported visibility in excess of twenty kilometers, a critical asset for facilitating their laser and GPS guidance systems. Pre-attack briefings included up-to-the-minute meteorological assessments ensuring that atmospheric interference would not degrade target acquisition or impact the precision of their guided munitions.

The F-16 squadrons coordinated their efforts via the revolutionary "Link 16" tactical data link, in combination with nearby AWACS and forward air controllers. Ironically, the IAF had now incorporated their own aerial drones into the fight, relaying battle damage assessments back to the strike group in real time, which allowed for immediate re-tasking of aircraft as necessary. One confirmed example involved the 117th Squadron, whose F-16s were redirected mid-sortie to a weapons cache in Deir al-Balah after drone imagery detected movement indicating an imminent rocket launch.

Pilot accounts, collected from mission debriefings, provide a gritty insight into the day's missions. One pilot from the 109th Squadron described the targeting protocol: "Coordinates were confirmed and authenticated on two separate systems before we received weapons release clearance. Our visual on the target was corroborated by drone footage piped directly to our cockpit screens." One of his comrades noted: "Synchronization was crucial—our attack run required precise separation between aircraft for maximum effectiveness and to avoid overlapping blast zones."

Strike outcomes were judged by pre-assigned metrics, including the destruction of critical command nodes and weapons stores. The IAF's post-strike analysis reported an effectiveness rate of 93%, figures that were later authenticated by third-party monitoring organizations. The Gaza War lasted barely three weeks, concluding in January 2009 with a unilateral Israeli ceasefire and a temporary reprieve from the Hamas rocket attacks.

UAVs & Unexpected Losses: 2010-2018

From 2010-2018, the IAF continued its longstanding vigilance in the skies over the Levant. As Israel's adversaries increasingly relied on low-signature aerial reconnaissance platforms, F-16s became first responders in a new and evolving battlefield: The fight for aerial supremacy against unmanned systems.

In December 2010, for example, an Israeli F-16 shot down an enemy spy balloon that appeared to be on a trajectory towards the Dimona Nuclear Research Center. Over the next six years, the IAF would shoot down three additional UAVs, all of which were Iranian-built and traced to Hezbollah. Each drone followed a similar pattern: Erratic, low-altitude flight paths intended to exploit natural terrain-masking and minimize radar exposure. Another intercept in April 2013 involved a drone that breached Israeli airspace from the northern coast near Haifa. This incident confirmed growing suspicions that Israel's adversaries were launching drones from offshore locations, thus bypassing ground-based launch risks.

Nevertheless, these UAV incursions prompted a major revision to the IAF's rapid response doctrine. Joint simulator drills and updated prioritization of aerial threats were gradually implemented into the IAF's training regimen. These events catalyzed a shift towards equipping F-16s with more advanced sensor pods and incorporating counter-UAV tactics into everyday readiness exercises, laying the foundations for a more robust aerial defense network.

On February 10, 2018, while continuing to counter the threat of enemy UAVs, the skies over the Levant erupted in the deadliest aerial conflict since the 1982 Lebanon War. The incident unfolded through a complex series of events: An Iranian drone incursion, retaliatory airstrikes from the IAF, and a barrage of Syrian anti-aircraft fire that brought down an Israeli F-16.

The crisis began around 4:30 AM, when an Iranian-built drone entered Israeli airspace from Syria near the Jordan Valley. Surprisingly, the drone was later identified as a duplicate American RQ-170 Sentinel, reverse-engineered from a sample captured by Iran in 2011. As it turned out, the drone had been launched from the Tiyas airbase in central Syria, a known hub for Iran's Islamic Revolutionary Guard Corps (IRGC).

The drone was intercepted and shot down by an Israeli AH-64 Apache helicopter. But because the RQ-170 "clone drone" had violated national airspace, Israel considered it an act of aggression from Syria. In immediate retaliation, the IAF launched an airstrike against the Tiyas airbase. Israeli F-16s formed the backbone of the strike package, flying from Ramat David airbase. By most accounts, the strike was a tactical success, destroying Iranian mobile command centers and most of their attendant drone capabilities.

However, Syrian air defense forces reacted aggressively. Utilizing their Soviet-era SAMs—including the SA-5 Gammon—the Syrians fired more than two dozen missiles at the Israeli F-16s as they exited Syrian airspace. During that withdrawal, however, one F-16 was struck by an SA-5.

Luckily, the pilot and navigator ejected over Israeli territory near Kibbutz, while their two-seater F-16 crashed into a field near Harduf in the Lower Galilee. Although both crewmen survived the ejection, both sustained injuries from the descent. They were later recovered by a heliborne rescue team and evacuated to Rambam Medical

Center in Haifa. It marked the first time in history that an Israeli F-16 had been downed by enemy fire, and Israel's first loss of a combat aircraft since the 1982 Lebanon War.

Following the loss of their F-16, Israel retaliated with a second, more extensive wave of airstrikes, targeting nearly a dozen Syrian and Iranian military installations within Syria proper. These included Syrian air defense batteries, Iranian command centers, weapons caches, and logistical nodes. This retaliatory airstrike showcased the IAF's ability to conduct deep-penetration strikes despite the threat of hostile air defenses. According to the IDF's battle damage assessment, this retaliatory wave destroyed nearly half of Syria's air defense network in under 90 minutes.

Apart from the tactical victory, however, the February 2018 incident had long-reaching strategic implications. It represented a turning point in Israel's campaign to counter Iranian activity in Syria. For the first time, the ongoing game of brinksmanship had entered a phase of direct fire confrontation against Iranian personnel in Syria. It also highlighted previously-undetected vulnerabilities in Israeli air tactics. Despite the F-16's high-tech defensive systems, the sheer volume and timing of Syrian SAM launches (combined with the relative predictability of the egress routes) proved sufficient to down an Israeli jet. The IAF's after-action reviews characterized the event as a tactical miscalculation. They concluded that the flight path during egress had exposed the F-16 to a SAM trap, and that the crew had not performed proper evasive maneuvers. Subsequent operational doctrines were adjusted to improve evasive tactics, timing, and coordination with AWACS.

Despite the aircraft loss, however, Israel had reasserted

aerial dominance with overwhelming retaliation. Still, this incident of February 2018 was a stark reminder that even advanced fighters like the F-16 were not invulnerable in the modern threat environment.

Israel-Palestine Crisis (2021)

In early 2021, a series of confrontations erupted at the Al-Aqsa Mosque compound in Jerusalem alongside a planned eviction of Palestinian families in the Sheikh Jarrah neighborhood. As violence surged, Hamas and the Palestinian Islamic Jihad (PIJ) launched more than 4,000 rockets into Israeli territory, targeting cities like Tel Aviv and Beersheba. In response, the IDF launched Operation *Guardian of the Walls* on May 10, which included a major aerial campaign over Palestinian territory—an operation wherein Israeli F-16s would play a central role.

As nightfall settled over the Mediterranean coast on May 13, 2021, the quiet skies above Gaza suddenly roared to life. In a coordinated aerial assault, Israeli F-16s surged forward in successive waves, unleashing a storm of precision-guided bombs and AGMs. Their target: Hamas' sprawling underground tunnel network—the so-called "Metro"—which they used for moving weapons and personnel beneath the urban terrain. Flying out of Ramon and Ramat David Airbases, these F-16s carried a combined arsenal of SPICE-1000 glide bombs, JDAMs, and GBU-39 Small Diameter Bombs. Using the proverbial "shock and awe" tactics, the F-16 squadrons struck from multiple vectors, executing the kind of joint operations that had come to define Israeli air power. Throughout the campaign, these F-16s dropped more than 450 precision-guided munitions, blasting tunnel entrances, collapsing underground depots, and destroying key Hamas command centers.

In many ways, the Hamas Metro operation highlighted a new era of Israeli air power: Surgically-precise air raids driven by cyclic, real-time intelligence and combined with massive force projection. It was the consolidation of all previous lessons learned over the past decade. F-16s conducted a number of simultaneous multi-axis attacks to overwhelm enemy air defenses and minimize warning time. Tight evasion maneuvers and tighter communication ensured that there wouldn't be a repeat of the February 2018 shootdown. IAF officials emphasized that real-time intelligence (much of it supplied by drones and signal intercepts) had been fused into the F-16s' mission planning system, which enabled dynamic targeting and re-targeting during the mission.

Throughout the 11-day conflict, F-16 strike fighters flew hundreds of combat sorties, often taking the lead for the IAF's ground-attack efforts. Apart from the keynote airstrike of May 13, Israeli F-16s flew a number of smaller surgical strikes against weapons-manufacturing plants, Command & Control centers, and rocket batteries.

Although Israel maintained that every target had been selected for its military value, international observers and news outlets took note of the civilian toll. Several buildings, including media offices belonging to Al Jazeera and the Associated Press, were allegedly destroyed by F-16 airstrikes. Israel stated these buildings were also being used by Hamas military intelligence, but the collateral damage nevertheless sparked global outrage and raised questions about the IAF's target verification process. For the F-16 pilots, however, the 2021 Israel-Palestine Crisis had been a familiar mission dressed in new complexities: Balancing firepower, accuracy, and global scrutiny in the unforgiving crucible of urban warfare.

The Israel-Palestine Crisis drew to a close on May 21, 2021, ending with an Egyptian-brokered ceasefire and a return to the *status quo ante*. Still, there can be little doubt that the conflict underscored the strategic salience of air power, especially in an asymmetric warfare environment. All told, the 2018 and 2021 operations highlighted the nuanced role of the F-16 in Israel's defense strategy. Beyond survival and tactical engagement, these incidents reflected a persistent adaptation to uphold air supremacy under the duress of rapidly-evolving threat profiles.

Ultimately, the Israeli F-16's combat history illustrates a narrative of resilience, adaptability, and strategic foresight. Each conflict underscored the multi-dimensional role this aircraft played—not just as a tool of warfare, but as a key actor in the geopolitical dynamics of a region where every operation carried wider implications. In that regard, the F-16 has built an enduring legacy of shaping tactical doctrines and geopolitical narratives.

Throughout its four decades of service, the F-16 has had a profound impact on the evolution of Israeli air power. From its inaugural victories in the Bekaa Valley to complex missions like Operation *Opera* and Operation *Peace for Galilee*, the F-16 has evolved to meet every threat, adapting through technological upgrades and strategic innovation. Their performance during more-recent operations like the Second Lebanon War and the 2018-2021 engagements showcased their overall adaptability, as well as their role in modernizing air combat tactics. As future conflicts loom on the horizon, the F-16 Fighting Falcon remains a symbol of Israeli air power and a critical component of its national defense doctrine.

Doha, Qatar: An ordnance trailer is parked in front of two American F-16Cs from the 614th Fighter Squadron, based at Torrejon Air Base, Spain. The F-16s are armed with bombs and AIM-9 Sidewinders as they prepare for another mission during Operation Desert Storm. To the immediate right of the F-16s are three F-1C Mirage fighters from the French Air Force. *US Air Force*

Chapter 3:
Desert Talons

On the morning of August 2, 1990, Iraqi forces under the command of Saddam Hussein invaded the emirate of Kuwait. Columns of tanks rolled through the oil-rich avenues, toppling the Kuwaiti government within a matter of hours. The world watched in shock, and then in outrage. The invasion drew fierce condemnation from the international community, prompting the UN to demand Saddam's withdrawal. But the Iraqi dictator didn't flinch. Instead, he massed his forces along the Saudi Arabian border and dared the world to stop him.

Saddam was certain that his army—the fourth-largest in the world and equipped with the latest in Soviet armor—would make short work of any rescue force that came to liberate Kuwait. He believed no Western coalition could stomach the cost of a high-intensity conflict, especially not the Americans. As he famously quipped, the United States was a "society that cannot accept 10,000 dead in one battle." Indeed, the memories of Vietnam were as galvanizing to Saddam Hussein as they were disheartening to the American public. He was confident that after the Americans had suffered a few thousand casualties, they would sue for peace on Iraq's terms.

But Saddam had miscalculated.
For as menacing as his army may have looked on paper, his air force was primitive by NATO standards. Iraq's MiG-25s and MiG-29s were hamstrung by pilots who lacked training and by command structures that prioritized optics over readiness. To make matters worse, Iraq's air defense

network was outdated and fragmented, a mistake that would cost them dearly in the opening days of the air campaign.

Saddam Hussein had clawed his way to power in 1968, riding the bloody wave of the Ba'ath Party revolution. From the chaos, he emerged as Iraq's strongman. By the early 1980s, he had gained absolute power. Statues rose in his image, streets bore his name, and he fancied himself as a modern-day Saladin or Hammurabi.

But then came the Iran-Iraq War.
It was supposed to be a quick victory against a wounded neighbor still reeling from the Islamic Revolution of 1979. Instead, it became a meat grinder: Eight years of trench warfare, artillery duels, perilous dogfights, and poisonous gas. The conflict ended in a bloody stalemate, claiming more than 300,000 Iraqi dead, and left Saddam saddled with a multibillion-dollar war debt, most of which had been financed by Kuwait. But rather than pay his debt to the Kuwaiti government, the "Butcher of Baghdad" decided to invade his neighbor to the south.

To justify the invasion, Saddam dusted off old grievances—including border disputes and oil rights. He alleged that the Kuwaitis had been slant-drilling Iraqi oil and that they were deliberately trying to lower oil prices by producing beyond OPEC's set quotas. Every accusation dripped with just enough truth to give it teeth. But the real motive was simpler: Kuwait held ten percent of the world's oil reserves and generated 97 billion barrels of crude each year. Thus, Saddam reasoned that if he couldn't pay his debt, he would simply annex Kuwait and take over its petroleum industry.

Thus, on the morning of August 2, more than 100,000 Iraqi troops and several hundred Iraqi tanks stormed across the border—the spearhead of an eighty-mile blitzkrieg into Kuwait City. Encountering only piecemeal resistance, Iraqi tanks thundered into the heart of the Kuwaiti capital, assaulting the city's central bank and carrying off with its wealth. Iraqi warplanes then strafed the Dasman Palace, turning the home of Kuwaiti ruler Emir Jabel al-Amhad al-Sabah into a smoking ruin. The emir and a few members of his staff barely escaped with their lives as they fled Kuwait by helicopter. The last transmission made over the state-run radio network was an appeal for help.

The UN responded with their normal variety of condemnations. Economic and military sanctions soon followed. But in Washington DC, something stirred. President George HW Bush authorized the first wave of American deployments to the region. Within days, the aircraft carriers Eisenhower and Saratoga were steaming towards the Persian Gulf while coalition fighter jets began pouring into Saudi Arabia by the hundreds.

This wave of deployments became known as "Operation Desert Shield"—a deterrent against Saddam Hussein lest he try to invade the Kingdom of Saud. However, when the skies of Arabia finally lit up in January 1991, Saddam's air force wouldn't just be outmatched.

It would be obliterated.

Throughout the conflict, the F-16 Fighting Falcon marked a powerful presence in the skies over Iraq and Kuwait. A total of 249 American F-16s deployed to the Gulf region during Operations Desert Shield & Desert Storm. These multirole fighters flew more than 13,500 sorties over the course of the air campaign, engaging in a variety of mission sets: Close Air Support (CAS), Suppression of

Enemy Air Defenses (SEAD), and strike missions against high-value targets deep inside Iraq. These missions highlighted both the reliability and adaptability of the aircraft. Operating in extreme conditions, and facing one of the most heavily-defended airspaces in modern history, the F-16 was central to the Allies' campaign to liberate Kuwait and destroy Iraq's military infrastructure.

The ordnance employed by F-16s during the Gulf War was as varied as their mission sets. Against hard targets and fortified positions, they used the Mk-82 and Mk-84 unguided bombs. When tasked with disabling runways or hardened aircraft shelters, F-16s often carried anti-runway cluster bombs like the CBU-87. For precision strikes, such as those against bridges or command bunkers, pilots employed laser-guided bombs, including the GBU-12 Paveway II. For SEAD missions, F-16s often carried the AGM-88 HARM air-to-ground missile. And, just as their Israeli counterparts had done in the 1980s, these American F-16s were armed with AIM-9 Sidewinders for aerial self-defense.

The operational tempo for Allied F-16 squadrons was relentless, often involving multiple sorties per day as crews cycled between mission planning, launch, and recovery. The range and payload of the F-16 gave it the ability to strike deep while its fly-by-wire reflexes made it a virtual ghost in hostile airspace, dodging threats and delivering precision hits with surgical finesse.

But what truly set the F-16 community apart was their relentless adaptability. Pilots and ground crews adopted new tactics, integrated advanced targeting pods, mastered night operations, and refined procedures for various inflight re-taskings. It was a level of innovation that set new standards for tactical air power. These advances, in turn, influenced the broader air campaign and strengthened the effectiveness of the overall war effort against Iraq.

Nevertheless, the achievements of the F-16 during Desert Shield and Desert Storm rested on the shoulders of a diverse but tightly-knit network of units. Squadrons drawn from different air wings across the United States, supported by maintenance, refueling, and logistical teams, worked in tandem to sustain this high operational tempo. Their collective efforts formed the backbone of coalition air power—laying the groundwork for the missions and personal acts of courage that would soon define the legend of American F-16s in the Gulf.

Desert Shield: Aug. 1990 – Jan. 1991

The 388th Fighter Wing was among the earliest F-16 units to arrive in the Gulf, deploying from Hill AFB, Utah. The wing brought with it a highly-trained cadre of pilots and ground personnel. Initial deployment began in August 1990, when the Wing's subordinate units—the 4th and 421st Fighter Squadrons—mobilized for overseas insertion. All elements and personnel moved quickly, supported by Air Force Logistics Command, which guaranteed that necessary supplies, ordnance, and support staff were in place to facilitate uninterrupted combat readiness throughout the deployment. This laser-focused efficiency was a significant logistical achievement, as it ensured the F-16s would maintain a high-surge capability across thousands of sorties. The 4th and 421st Fighter Squadrons touched down in the Arabian Peninsula almost simultaneously, arriving at Al Minhad Air Base in the United Arab Emirates (UAE) between late August and early September 1990. The timing of their arrival allowed for continuous training with coalition partners from the RAF and French Air Force—exercises that would prepare the 388th Fighter Wing for the intense operational demands of the upcoming air campaign.

Captain Arthur Gatti, a maintenance officer supporting the 4th Fighter Squadron described the early months of Desert Shield as a struggle to stay busy and find purpose in the broader mission. After completing his one-year remote tour in Korea, Gatti had just settled into his new post with the 388th Fighter Wing when Iraq invaded Kuwait. "I expected a hot and tedious 90-day show-of-force," he said. "I really didn't expect the Iraqis to persist when a large US-led contingent arrived in-theater."

As the assistant Officer-in-Charge of the 4th Aircraft Maintenance Unit (AMU), Gatti was responsible for the night shift, keeping twenty-four F-16C *Block 40* Fighting Falcons ready for combat in the blistering desert. But the dry heat wasn't just a nuisance, it was a threat. Dehydration was rampant, and Gatti's first mission was clear: Keep the troops hydrated, focused, and sharp. "The deployment seemed more like a punishment than a mission of purpose," he later admitted. But within 24 hours of their arrival at Al Minhad, the 4th AMU had every plane in the squadron ready to fly. "We spooled up to a normal daily flying training schedule and had settled into a routine within one week."

Meanwhile, two other F-16 units, the 17th and 33rd Fighter Squadrons (363rd Fighter Wing) deployed from Shaw AFB, South Carolina to Al Dhafra Air Base, UAE, beginning on August 8, 1990—only six days after the invasion of Kuwait. Respectively, the two squadrons deployed with their *Block 25* F-16C Fighting Falcons.

Personnel and aircraft from Shaw AFB departed via C-5 and C-141 transports, while the F-16s made their transatlantic journey with multiple inflight refuels. Within days of their arrival, both squadrons assumed CAP station duties to defend the regional airspace. Together, their presence became a critical component of US air power during the early days of Desert Shield—training, patrolling, and maintaining deterrence throughout the fall of 1990.

On August 15, the 70th Fighter Squadron, stationed at Moody AFB, Georgia, began its deployment to Saudi Arabia. Their movement involved twenty-four F-16C Fighting Falcons and more than 500 personnel. Like their comrades from Shaw AFB, the 70th Squadron's aircraft flew across the Atlantic and Europe with air-to-air refueling support. Upon arrival, the squadron immediately began preparing for combat operations. The unit quickly established a high sortie rate and began flying defensive air patrols over Saudi territory. Their mission (as part of the broader Desert Shield outlook) was to protect Saudi airspace against any potential incursion from the Iraqi Air Force.

At around the same time, the 401st Air Expeditionary Group at Torrejon Air Base, Spain, deployed two of its F-16 squadrons: The 612th and 614th. The 612th Fighter Squadron deployed to Incirlik Air Base, Turkey, while the 614th Fighter Squadron established a forward-operating base in Doha, Qatar, where it became the core of the 401st Fighter Wing (Provisional), formed from the Air Expeditionary Group's support elements.

During these opening months of Desert Shield, US Central Command Air Forces (CENTAF) kept its tactical air squadrons on a tight leash, imposing strict parameters for all in-theater sorties. As the military coalition grew, Lieutenant General Chuck Horner (USAF) became the commander of Allied air forces. More than 2,500 tactical aircraft would be under his command—1,800 of which were American. It was, by all measures, the largest aggregate air force since World War II.

Meanwhile, CENTAF and coalition leaders created the Tactical Air Command Center (TACC) to plan and organize the anticipated air campaign against Iraq. The forthcoming tactics, techniques, and procedures were eventually codified in a 600-page document known as the "Air Tasking Order" (ATO). Within its pages, the ATO included maps of aerial routes, lists of anticipated targets, and target arrival times. Additionally, the ATO assigned specific missions, targets, ordnance loads, and optimal routes for the various F-16 squadrons in theater. According to TACC guidelines, the F-16's mission sets would be tailored according to each squadron's configuration. For example, CAP missions would be prioritized according to which F-16 squadrons had *Block 30* variants as opposed to the *Block 40* series.

By November 1990, the skies over Saudi Arabia were humming with the arrival of incoming warplanes. Coalition air power was swelling by the day—fighters, bombers, tankers, and watchful AWACS lining up like sentinels across the sun-bleached tarmacs. That same month, the UN Security Council passed Resolution 678, drawing a hard line in the sand: Saddam Hussein had until January 15 to pull his forces out of Kuwait...or face annihilation.

Still, the Iraqi dictator showed no signs of backing down.

On paper, the Iraqi Air Force was outgunned and

outnumbered by a modern, multinational air arm. But Saddam still had more than 500 Soviet-built aircraft, and his pilots weren't green. Many had survived the bloody crucible of the Iran-Iraq War. They knew the terrain; they knew the stakes. And when the time came, they'd be ready to die defending the skies over Iraq.

On December 28, 1990, the growing contingent of CENTAF F-16s were joined by the 10th Fighter Squadron, part of the 50th Fighter Wing from Hahn Air Base in Germany. Sending their twenty-four F-16Cs to Al Dhafra Air Base, the pilots and ground crews were attached to the 363rd Fighter Wing upon arrival, integrating with the 17th and 33rd Squadrons already present.

Said Captain Bill Andrews, an F-16C pilot with the 10th Fighter Squadron:

"An interesting thing about deploying out of USAFE [United States Air Forces Europe] is that we were organized and prepared to fight in place. All of our bases in Europe are well-equipped with aircraft shelters, squadron shelters, etc. Our maintenance was designed to fight right out of our backyard, and we didn't have 'deployment plans' like the units from the States. The war was supposed to be on our turf."

As such, European-based units like the 10th Fighter Squadron had to learn the logistics of becoming an "expeditionary force" overnight. In fact, Andrews recalled that his squadron was so uninformed of their anticipated expeditionary requirements, that they sent an advance party to Al Dhafra just to "figure out exactly what we needed to bring."

The following week, two US Air National Guard squadrons were mobilized to supplement the growing need for air power. The 138th Fighter Squadron (New

York Air National Guard) deployed to the Persian Gulf on January 6, 1991. Equipped with the slightly older F-16A/B variants, the squadron deployed a total of twelve aircraft and approximately 400 personnel to Al Kharj Air Base in Saudi Arabia. Though originally intended for air defense missions, the squadron's F-16s were quickly modified to perform precision ground-attack missions, receiving new software and hardware upgrades to accommodate the incoming precision-guided weapons.

The other "citizen-airman" unit—South Carolina's 157th Fighter Squadron—likewise arrived at Al Kharj Air Base in January 1991. Just like the 138th, their service demonstrated the seamless integration of Guard and Reserve units within the active-duty framework. These National Guard F-16s expanded the coalition's operational footprint and would prove invaluable during the high-intensity phases of the air war.

Desert Storm: Jan-Feb. 1991

It came as no surprise when, on January 15, Saddam Hussein missed his deadline and refused to back down. Instead, his answer to diplomacy was silence and steel. The Butcher of Baghdad wasn't going anywhere. The following day, President Bush addressed the nation, announcing that the long wait was over.

Operation Desert Shield had just become Operation Desert Storm.

On January 17, at 2:38 AM, Baghdad time, the first wave of the coalition's air campaign destroyed Iraqi radar sites near the Saudi border. Coalition warplanes swept in low and fast, punching through Iraqi airspace like a thief in the night. Within minutes, the "eyes and ears" of Saddam's air defenses went dark. The war had begun, not with a massive ground invasion, but with precision and shock. Baghdad was about to feel the full weight of modern air power. And the F-16 Fighting Falcon would be at the forward edge.

Assignment patterns were not arbitrary; they reflected the ongoing reassessments being made by CENTAF, based on feedback from mission results and real-time intelligence. Over the course of the air war, F-16 units would be moved, tasked, or re-tasked based on enemy movements and post-strike analyses. This fluid criteria, backed by a strong maintenance program, allowed F-16s to stay at the forefront of combat, directly impacting the pace and effectiveness of coalition advances.

Many squadrons faced unique challenges throughout the air campaign, particularly those fighting within the theater's busiest airspace sectors. From the "early bird" F-16s of the 388th Fighter Wing to the late-arriving Vipers of the Air National Guard, each unit developed its own reputation for resilience and adaptability under pressure. Their mission histories set the stage for individual pilots and crews whose experiences provide deeper insight into the F-16's legacy in Desert Storm.

Opening Day

In the predawn hours of January 17, 1991, the wail of jet engines cut through the air at Al Minhad Air Base, UAE. The long-anticipated air war against Iraq had begun, and the 388th Fighter Wing would be at the tip of the spear. Its two primary squadrons, the 4th and 421st Fighter Squadrons, were flying F-16Cs equipped with the revolutionary LANTIRN targeting system. These *Block 40* aircraft gave pilots the unprecedented ability to strike targets at night or in poor weather, while flying at low altitudes, and deliver laser-guided bombs with surgical precision.

In the days leading up to Desert Storm, the ominous silence that hung over the hardened shelters at Al Minhad was punctuated by the tapping of maintenance tools and the muffled conversations of aircrews verifying every checklist. Faces revealed determination and restraint, but tension hung heavily over the proceedings. Maintenance crews tightened harnesses and fueled aircraft while sun-weathered hands checked over the vital components of the armament suite and life support systems. Each action revealed a singular focus: to ensure that man and machine would be ready to join the largest air campaign since World War II.

Unwittingly or not, the 388th Fighter Wing shouldered a significant share in the opening phase of Desert Storm. Its primary mission, as mandated by CENTAF, was to suppress enemy air defenses and disable the Iraqi Command & Control network within the first few minutes of the conflict. F-16s from the 4th and 421st Fighter Squadrons held a dual responsibility: Striking radar installations, SAM sites, and communications nodes while holding fast to the principles of operational tempo. Their mission was to pierce the veil of Iraq's air defenses, allowing follow-on strike packages to reach deep into enemy territory. For these inaugural missions, CENTAF prioritized precision over brute force, aiming to cripple Saddam's ability to coordinate responses during the early hours of the air campaign.

Within the Wing's dynamic structure, the 4th and 421st Fighter Squadrons operated as tightly-knit units, each bringing distinct experience and equipment sets to the mission. The 4th Fighter Squadron specialized in low-level ingress profiles and precision-guided munitions delivery. The 421st Fighter Squadron, with its own cadre of seasoned pilots, focused on suppressing enemy radar threats and engaging secondary targets as needed. Specialized EW equipment and chaff dispensers lined their aircraft, offering a critical edge against both radar-directed anti-aircraft guns and SAM batteries that dotted the approaches into Baghdad.

Final mission briefings started before midnight in shielded, sandbagged rooms. Cellophane overlays, target photographs, and complex maps covered the tables while mission planners recited route timings, entry points, and threat envelopes. The squadron commanders moved through the room, their voices low but firm, reviewing rules of engagement and reinforcing the chain of command that would govern every radio call and in-flight decision. Flight leads and their wingmen pored over data on enemy air defense radars west of Basra and anticipated the activation of various SAM sites (including the SA-3 and SA-6 missile launchers). With the lingering possibility of chemical warfare, pilots and crews reviewed their emergency protocols and countermeasures, rehearsing every step to evade (or survive) a contaminated environment. Each pilot took mental inventory of EW techniques, including jamming frequencies and the deployment of anti-radiation missiles designed to track enemy radio emissions. Support personnel oversaw ordnance loading and confirmed fuel reserves, marking hardcopy manifests and coordinating launch windows with the broader ATO.

In the final hours before launch, there was a nagging

apprehension of the unknown. Intelligence indicated that Iraqi air defense systems had been networked into a dense and unpredictable array. Iraqi MiGs stood ready on their runways, while tracking radars swept the horizon. Still, the Allied pilots and ground crews kept themselves busy, checking and rechecking wing-mounted pods, and confirming the status of their onboard jammers. Others exchanged brief nods, acknowledging the unspoken risks ahead. Airmen moved purposefully between rows of F-16s. Every checklist, every inspection, and every final handshake underscored the seriousness of the moment.

Shortly before 3:00 AM, the first wave of F-117 Nighthawks began destroying Iraqi air defense radars and command centers with their precision-guided bombs. At 3:03 AM, the 421st Fighter Squadron launched eight of its own Vipers into the early morning gloom. They were among the first conventional fighters to enter Iraqi airspace. Their mission: neutralize radar installations, communication hubs, and early-warning systems.

Flying low and fast, the 421st's F-16s made use of the LANTIRN system to navigate terrain and identify targets through pitch-black darkness. This was a significant evolution in air combat tactics. Previously, such missions were confined to daylight hours or required extensive SEAD support. But the 421st Vipers were part of a new breed: Multi-role aircraft able to independently find, fix, and finish targets. Each F-16 carried a heavy load: a pair of 2,000-pound Mk-84 bombs, two AIM-9 Sidewinders, and external fuel tanks for the long haul. Pilots refueled twice along the route from Al Minhad to southern Iraq, rendezvousing with KC-135 aerial tankers over the Persian Gulf.

At first, the F-16s encountered turbulent weather over

the Gulf, and nearly had to abort the mission, but things started to clear up nearly as soon as they reached the mainland. As Captain Mark Miller recalled, the 421st's ingress route was saturated with anti-aircraft fire. "I saw two SA-6s," he recalled. "One was tracking on me, but it detonated early. I saw flashes of light through the clouds. The ZSU-23-4 looked like a fire hose of red tracers. We were out of their range, though. And since our lights were off, they never really came close to hitting any of us."

Captain Michael O'Grady painted a similar picture from the view of his cockpit. As his own F-16 crossed the Iraqi border, he thought to himself:

"Get ready, the fireworks are about to start."

Forty miles from the target, his wingman suddenly called out the first of several SAMs fired at their formation. "For the next five minutes, the string of F-16s reacted to an extremely dense network of SAM and anti-aircraft fire," he said. "This was, no doubt, the *longest* five minutes of my life."

The Vipers broke formation in disciplined chaos, defensive flares billowing out, wings rattling through the turbulence of evasive maneuvers.

Then another SAM. And another.

The sky was ablaze with streaking missile trails and the angry bark of anti-aircraft fire. For those five long minutes, the beleaguered F-16s ran the gauntlet through one of the most brutal air defense networks on earth.

Still, O'Grady pressed on. "I rolled in on the target," he said, "released the 4,000 pounds of high-explosive bombs [double Mk-84s] on my aim point, then watched my wingman do the same." The twin 2,000-pound bombs dropped clean off the rails, plummeting towards the enemy target below. O'Grady rolled out, punched the throttle, and banked hard to the west. Behind him, his wingman called out, "Good splash." Then came the secondary explosions—bright flashes erupting across the target zone.

Egress was fast and low, their adrenaline masking the onset of exhaustion. However, the tension eased with every passing mile back towards the Kuwaiti border. "After crossing the border," he recalled, "I felt a sense of euphoria, accomplishment, and relief." Cheers erupted over the radio net, tight and controlled, but infused with the unmistakable timbre of relief. No one was missing. No one had been shot down. Against all odds, every pilot had survived. "As I coasted back to home base," he continued, "I felt very proud of myself and my squadron."

By 6:00 AM, all eight F-16s were on final approach to Al Minhad. Eight planes had departed; all eight had returned. Awaiting their arrival, nearly every pilot and ground crewman in the 388th had crowded onto the taxiway. "As each aircraft touched down, another came into view," said Captain Keith Rosenkranz, another F-16 pilot in the 421st. "When the eighth light emerged from the morning haze, everyone began to cheer. When the pilots finished debriefing maintenance on the status of their jets, we escorted them back to the squadron. Everyone was on an emotional high. We were expecting to lose two, maybe three pilots. But every one of them made it back safely."

Meanwhile, as the morning sunlight crept through the haze, the 4th Fighter Squadron prepared for its daylight mission. Their target set was equally ambitious: The backbone of the Iraqi war machine itself—Republican Guard command centers, communications nodes, and logistics hubs. Although the squadron had trained extensively for night operations, their mission today would be a broad daylight strike into heavily defended territory, targeting enemy infrastructure that had survived the initial stealth raids.

Pilots reviewed the day's flight plan and intelligence updates. The broader mission called for coordinated strikes on southern Iraq, with the goal of decapitating Saddam's command structure in the region. Before the first plane took off, however, Arthur Gatti and his 90-man maintenance crew had worked feverishly throughout the night, preparing the squadron for action. Every mechanic, weapons loader, and avionics technician knew what was at stake. That night, twenty prime F-16s and four "spares" (backup aircraft) had been loaded and prepped for combat. The hangars echoed with the clang of tools and bellowing commands as adrenaline replaced fatigue. Gatti himself walked the flightline just before dawn, inspecting jets, chatting with pilots, and ensuring no detail went unchecked. He shook each pilot's hand, offering reassurance: "You're taking the best jets in the Air Force into combat."

Each F-16 launched with a pair of Mk-84 bombs slung under its wings, backed by a full load of chaff and flares to counter the threat of Iraqi SAMs. The enemy's air defenses, though battered by the early morning raids, remained dangerous.

Flying in at low altitude and under radio silence, the F-16s crossed the Kuwaiti border and pushed northward. On radar, the Iraqi defenses began to stir. Ground-based radars swept through the sky. Missile tracking systems activated. Within moments, the Vipers began to register enemy lock-ons.

The battlefield was live.

Undaunted, the F-16s of the 4th Fighter Squadron pushed forward, where they executed one of the first large-scale daytime strikes by non-stealth aircraft during the war. Beneath them lay the target zone: hardened bunkers, Command & Control facilities, microwave relay towers, and suspected HQ bunkers used by elite divisions of the Republican Guard. Dropping from medium altitude, the F-16s released their payloads with surgical precision. Explosions rippled across the landscape, leaving twisted wreckages and pillars of smoke in their wake. A few enemy SAMs reached skyward, but none found their mark. The F-16s soon cleared the area, returning south through an egress corridor now marked by combat.

Over the radio, strike leads confirmed direct hits on multiple targets. Satellite imagery and AWACS feedback later confirmed that several command bunkers and communication hubs had been destroyed. By late afternoon, the Vipers began arriving back at Al Minhad, the last one touching down just after 4:30 PM, with heat shimmering off the tarmac. The flight line soon populated with exuberant (albeit exhausted) ground crews, inspecting battle damage and rearming aircraft for what would be another long night ahead. Captain Gatti walked the line again, debriefing aircrews, checking on maintenance teams, and scribbling reports for minor issues caught during recovery. Despite fatigue and the swirling dust, there was pride in every handshake and backslap.

Gatti then flagged down one of his pilot friends, Lieutenant Daniel Swayne. As Gatti recalled: "He recounted the highlights of his mission and how pleased he was with his jet and its performance. I've never felt prouder of the men and women of the 4th AMU than at that moment." For the pilots and ground crews in the 388th Fighter Wing, however, the day had been a test of everything they'd trained for: A full-spectrum strike against ground targets defended by some of the densest SAM coverage in Iraq.

And every aircraft in the unit had returned.

Behind the scenes, analysts tallied results. The 421st had crippled a number of radar and communication nodes during the pre-dawn strikes. Meanwhile, the 4th had hammered away at the Republican Guard's Command & Control capabilities, contributing to the broader disruption of Iraq's military cohesion. The opening day of Desert Storm had been a proving ground for the modern F-16 and the operational concepts built around it. For the 388th Fighter Wing, it was a day that validated months of preparation, innovation, and resolve. The 388th had not only delivered a decisive blow against Iraqi forces, they had set the standard for the rest of the air campaign.

As twilight descended onto the horizon, another shift of pilots and ground crews prepared for another night of war. But for those who had flown (and for the maintenance crews who had launched them) January 17, 1991 was a day they would never forget. The F-16 Fighting Falcon had drawn first blood in a new kind of air war—fast, precise, and unrelenting.

In the coming days, the operational tempo surged as Allied forces continued hammering away at key targets in Iraq and Kuwait. Iraqi defenses, while degraded, were still dangerous. SAM launchers and anti-aircraft fire remained a persistent threat, especially as F-16s crossed into more-contested zones near Baghdad. Still, the significance of the 388th's operations on Day One cannot be overstated. Coalition aircraft launched more than 1,200 sorties during the first fourteen hours of Desert Storm. Among these, the F-16s flying from Al Minhad were among the most active and effective. A combination of superior training, superior technology, and aggressive tactics had allowed the 388th to punch well above its weight class.

As the sun set over Al Minhad and the Arabian Peninsula, there was no celebration. Just the hum of maintenance trucks, the scent of jet fuel, and the quiet knowledge among the airmen of the 388th that the war had begun...and that they were leading the charge.

Package Q Airstrike

By January 19, the skies over Iraq were raining fire. In just 48 hours, the air campaign of Desert Storm had opened with a vengeance—thousands of coalition sorties tearing into Saddam's defenses with merciless precision. Early-warning radars had been gutted, air bases cratered, and command nodes shattered. There was no question: Iraq's air defense network was unraveling at the seams. Meanwhile, the Iraqi Air Force, battered and hesitant, had lost nearly a dozen fighters in the opening rounds of aerial combat. Several other aircraft had been destroyed on the ground, or lay hidden behind hardened shelters, effectively ceding control of the skies to the UN Coalition.

Still, Baghdad itself remained a forbidding fortress. Anti-aircraft guns and radar-guided SAMs ringed the city. To this point, only the faceless, radar-evading F-117 Nighthawks had dared to breach the capital's air defenses, striking the regime's power centers under the cloak of night.

But that wasn't enough.

Baghdad still stood, and Saddam's regime still claimed

defiance. Thus, CENTAF planners drew up something bigger, something more audacious: A massive airstrike codenamed "Package Q." It would be the largest daylight raid of the war to date. Dozens of aircraft flying against dozens of targets, it was a hammer blow aimed straight at the nerve center of Saddam's regime. The mission had three objectives: Shatter high-value assets in downtown Baghdad, send a psychological shockwave through Saddam's inner circle, and prove to every Iraqi citizen that nowhere—not even the heart of their capital city—was safe from Allied air power.

Package Q was an airstrike on a scale the Iraqis had never seen. It would become the largest operational F-16 strike in history—a full-force show of steel and fury aimed straight at the heart of Baghdad. Seventy-two Vipers were tapped for the mission: Fifty-six from the 388th Fighter Wing, with an additional sixteen from the 401st Fighter Wing, each one loaded with a pair of 2,000-pound bombs.

But they wouldn't go in alone.

Eight F-15 Eagles would provide fighter escort, ready to tear apart any MiG that dared lifting off. Eight F-4G Wild Weasel jets armed with HARMs and a pair of EF-111 Ravens would provide radar suppression, jamming enemy scopes to give the strike package a fighting chance. This wasn't just a mission. It was a coordinated symphony of destruction.

And the choreography was bold. Jets would launch from four separate airbases—different countries, different clocks—and thread the needle of timing to rendezvous in the skies above Saudi Arabia. From there, they'd swing west toward Syria, a feint to mislead the Iraqi command into thinking the mission was a Scud-hunting sweep. Then came the pivot: A hard break eastward and a high-speed plunge into Baghdad's kill zone.

In this high-stakes calculus of the air campaign, a daylight mission offered both risk and opportunity. For the pilots, the imminent threat came from one of the most densely-layered anti-aircraft environments ever encountered. Air Intelligence briefings underscored the high probability of attrition. The defenses around Baghdad were robust, centrally-coordinated, and battle-tested. Major Jeff Tice, a flight leader in the 614th Fighter Squadron, reflected on the inherent danger: "Baghdad was the most heavily-defended piece of real estate in the world at the time." Still, the coalition was eager to show that even Baghdad was not invulnerable.

The pilots of the 388th and 401st Fighter Wings knew the risks. They'd seen the maps and read the intel reports. Baghdad was a citadel, and they were being sent straight into the lion's den. But bravado ran deep in the blood of these fighter pilots, and on that hot desert morning, confidence masked the fear. In the 614th Squadron's briefing room, Jeff Tice listened intently, eyes locked on the mission board. The air was thick—not with talk, but with tension. No one said it out loud, but they all felt it: *This mission was different.* Still, in the minutes before engine start, they reassured each other with tight grins and shoulder pats. "It won't be that bad," they told themselves. Tice would later admit, "We honestly thought it wouldn't be as rough as it looked on paper."

But paper doesn't shoot back.

Meanwhile, at the Al Minhad Air Base, deep inside the briefing cell of the 421st Fighter Squadron, the room was dead silent except for the low murmur of shuffling paper and the occasional clicking pen. Keith Rosenkranz sat silently among his fellow pilots, absorbing the details with a mixture of determination and silent calculation. He had reviewed the anticipated threat environment, noting the locations of Iraqi SAMs and air defense guns. Every red circle was a death zone, and every missile symbol a reminder that Baghdad's airspace was the most dangerous in the world.

The prize target for today's mission: Al Tuwaitha Nuclear Research Center.

This wasn't just another bombing mission; it was history repeating itself. It was a chance to finish what others had started. The Osirak reactor, destroyed by Israeli F-16s a decade earlier, had once been the centerpiece of Saddam's nuclear ambitions. But Iraq had been rebuilding parts of the facility, and was suspected of continuing clandestine research for atomic weapons. That made Al Tuwaitha more than just a military target. It was a symbol of unfinished business and a stark reminder of Saddam's perennial defiance.

That morning, dozens of F-16s sat ready on the ramps at their bases throughout the Gulf. Each Viper wore a pair of 2,000-pound bombs under its belly, missiles on its wingtips, and extra fuel tanks filled to capacity. The pilots, meanwhile, moved with a methodical sense of urgency—consulting checklists, reviewing the flight plan, and double-checking their maps. The mission's route to Baghdad required exact navigation through pre-identified corridors, skirting the strongest concentrations of enemy SAMs, but never fully escaping that threat. Flight leads coordinated the timing of refueling points and contingency procedures should any aircraft suffer damage or malfunction en route to Baghdad.

As the sun rose, the skies over the Persian Gulf sprang

to life with the steel and fury of American air power. Seventy-two Fighting Falcons, the largest F-16 strike package ever assembled, surged into the pale sunlight, the roar of afterburners rolling like thunder. They climbed westward, refueling over Saudi Arabia, with each mile bringing them closer to the cauldron that was Baghdad.

As the Vipers of Package Q closed in on the Iraqi capital, the sky below them erupted into chaos. "The world lit up underneath us," Jeff Tice recalled—and he wasn't exaggerating. In an instant, the skies over Baghdad became a firestorm of light and steel. Anti-aircraft guns thundered from rooftops and bunkers, their tracers lacing the sky like a flaming spiderweb. Flak bursts filled the air, creating a cloud cover all their own. One pilot likened it to "seeing an overcast day" as he flew into the target zone.

The Iraqis hadn't expected this massive, daylight punch into the heart of their capital city, and they panicked accordingly. "They were just throwing AAA and SAMs into the sky…hoping for a hit." The F-4G Wild Weasels, tasked with neutralizing the worst of the enemy radars, fired off their HARMs at some outlying air defenses, but fuel constraints had forced them to peel away before the main strike force hit the target area. And, just like that, the F-16s were alone—charging into Baghdad's defenses without any SEAD support. Missile warnings screamed through cockpits, and radios filled with panicked voices calling for evasive turns.

In the middle of it was Major Emmett Tullia, flying the F-16 call-signed "Stroke 3," with the 614th Tactical Fighter Squadron. He led his formation starboard, guiding them toward an oil refinery straddling the Euphrates, when his HUD lit up like a Christmas tree.

He didn't need to guess.

A pair of SA-2s tore upward from the earth—long white fingers of death clawing at the sky. As Tullia recalled, "I turned around, saw them coming up. They went beneath us and overshot." Undeterred, he rolled in and pickled his bombs, watching them fall squarely onto the refinery. Fire and smoke ripped through the structure, sending the oil-fed flames boiling skyward. But there was no time to celebrate.

Another threat warning: Two more missiles.

Tullia turned south to escape—but behind him, another pair of missile plumes boiled into the air.

"Oh, no, time to start maneuvering,'" he thought.

He jettisoned his wing tanks and dove hard to break their lock. Just then, another pair of SAMs slashed their way into the sky, narrowly missing his F-16. Their contrails split the air just yards from his canopy.

And then, three more.

They came in from different directions, moving faster and smarter. Over the radio, his voice cut through the chaos:

"Stroke 3, defending 6!"

Six missiles. All hunting him.

Now in survival mode, Tullia was twisting his jet through impossible angles, his F-16 bleeding speed with every desperate turn. He later admitted, "It was challenging because I didn't have a lot of chance to gain airspeed." He was low, slow, and dangerously close to stalling out. One SA-6 missile passed so close to his cockpit that he could hear the hiss of its rocket motor.

With altitude and speed vanishing fast, Tullia made the only choice he had left: he dove straight into the hornet's nest of anti-aircraft fire. Dozens of Iraqi gunners below locked onto the plunging F-16. Puffs of flak erupted all around him, but the dive worked. The Falcon picked up speed, bit into the air, and at the last moment, Tullia pulled hard, clawing his way back into the sky.

He was alive, but just barely.

By the time he coasted into his final approach at the air base in Qatar, his jet was running on fumes. His ground crew quickly swarmed the aircraft. Only then did they realize that *none* of his flares or chaff dispensers had worked; he had evaded *six missiles without any countermeasures.*

For his courage, precision, and sheer survival instincts under fire, Major Emmett Tullia was awarded the Distinguished Flying Cross. But medals aside, he had ridden the edge of death and had come back with his jet (and his life) intact—a feat that no training mission could have prepared him for, and no simulation could have replicated.

Meanwhile, in the 421st Squadron's sector of the battlespace, Keith Rosenkranz narrowed his focus onto the Al Tuwaitha facility. His flight's ingress into Iraq via Saudi airspace had been eerily quiet—too quiet, in fact. No radar locks. No missile trails.

"Picture clear," the AWACS had said.

But clear didn't mean safe. Either way, he didn't trust it.

Below him, the desert shimmered in muted tans and blood oranges, broken only by the tarry roads and the mighty Tigris River, winding past the crumbling ribs of old airfields and command bunkers. The release point for their bombing run was Shayka Mazhar Airfield, a ghost of a base that once bristled with Iraqi fighters, now just another cratered relic. Once they passed Shayka Mazhar, the 421st would be clear to initiate their attack on Al Tuwaitha.

The shimmering flows of the Tigris provided a perfect path to the target: Al-Tuwaitha was only eighteen miles north, prominently nestled along the eastern crook of the river. The afternoon sun, now dropping towards the west horizon, worked in their favor. Iraqi gunners would have to squint into its blinding fury to find the prowling F-16s.

Twenty miles south of Shayka Mazhar, the silence finally broke. A cold buzz hissed inside Rosenkranz's headset, dull at first, then rising into a scream. On his threat receiver, a crimson SA-2 symbol lit up. The Iraqi radar had locked onto him. And before he could say a word, the missile was in the air.

"Collar 16, launch right two o'clock!"

That was when his training took over. Stick in hand, he broke into a hard weave, jerking the Viper through evasive maneuvers while shooting off thick clouds of disruptive chaff. The world blurred. His radar chirped again: Another missile, straight ahead.

"Collar 16, two launch on the nose!"

No time to panic. He kept weaving, eyes slicing across the sky, until there it was: A white finger of smoke, climbing fast and hungry, headed straight for him.

"Collar 16, tally, cons on the nose!" he cried out.

Shayka Mazhar flashed below, forgotten. All around him, the threat receiver was going ballistic. "SA-2s, SA-3s, SA-6s, SA-8s—they were everywhere!"

The sky lit up with flashes of orange and yellow, erupting from both sides of the Tigris River. Iraqi SAM sites were firing off with a vengeance. Missiles clawed their way skyward, trailed by the skeletal fingers of white and gray smoke. One missile streaked past his left wingtip, another split the sky to his right. Then came the flak. "Popcorn-shaped cloudbursts" as he called them—angry black clouds that blossomed in all directions. The air had become a firestorm of shrapnel and chaos.

Ahead and slightly to his left, Captain George Stillman, one of his wingmen, held position about 2,000 feet away. Rosenkranz pulled hard right to maintain formation, dodging fresh contrails as SAMs kept rising like javelins from Hell. The radios were swamped. Callouts were buried beneath the static and shouted warnings.

By now, a new pair of SAMs had locked onto their formation. Time was vanishing; and their options were shrinking. If the Vipers didn't act soon, they'd be nothing but smoke trails and wreckage within a few moments. Suddenly, one of the missiles detonated behind him, low at his left eight o'clock position. The missile gave a brilliant burst, but it was too far off to have any effect on his plane. The other missile kept coming, but it appeared to be running out of steam.

With baited breath, Rosenkranz stared it down.

Then, mercifully, the SAM faltered, arcing back towards the ground like a defeated predator.

Shaking off this close encounter, Rosenkranz rechecked his HUD: Thirteen miles to Al Tuwaitha. Flipping the switch, he called up "Air-to-Ground Mode" and dropped altitude to pick up speed—one eye on the river, the other eye on Stillman. Flak erupted around them like a field of black blossoms, blooming and dying in seconds. "Thicker than soup," he'd later recall. The sky wasn't blue anymore. It was a dismal steel gray filled with smoke, fire, and death.

At nearly eight miles from his target, another missile signal popped onto his scope.

Then another.

And another.

His radar receiver screamed like a wounded animal. "The Iraqi gunners know exactly where I'm at," he would later say. "And they have me locked." Launch tones stacked one after another.

He couldn't process it anymore.

"I hear the radio. I hear the tones. But I can't think about what they mean. All that matters is hitting the target. If I get shot down, then so be it."

He scanned the satellite photo, then glanced outside looking for the river bend, the telltale curve that marked the approach. Al-Tuwaitha should be just to the right. Stillman was already rolling in. Rosenkranz took a beat...then followed.

He rolled hard into the bank, slicing the horizon at ninety degrees. The Viper dropped like a blade. Passing through 22,000 feet, he got the max-toss cue in his HUD. But he still couldn't see the target. He rolled steeper to 135 degrees, when, finally, the diamond target indicator appeared in the HUD. But the complex was shrouded in smoke. Coalition bombs had already hit parts of the facility, clouding the aim point in ash and debris. He had no visual, but at this point, his only option was to hit the "pickle switch" on the diamond.

The jet shuddered at 14,700 feet, the weight of the bombs releasing beneath him. The aircraft kicked upward into a 5G recovery climb, nose clawing for the horizon. Flak bursts pounded the sky like angry fists from below, shaking the cockpit. Rosenkranz shoved the throttle into afterburner, the engine roaring like a dragon. "I'm in the heart of the SAM envelope," he later said. "If I don't gain some altitude in a hurry, life as I know it will come to an end."

And at that moment, it almost did.

He looked to his right and saw a bright red object—a missile—streaking towards him from his nine o'clock position. He yanked the stick back with everything he had.

"Collar 16, SAM launch, right nine!" he screamed.

The jet moaned in protest, bleeding energy fast. Passing 20,500 feet, his speed had dropped below 200 knots. Deadly slow and vulnerable.

"If I'm going to outmaneuver this missile, I have to decrease drag and regain my energy."

At this point, his only hope was to jettison his external fuel tanks. He slammed the emergency jettison switch. The Viper jolted as its wing tanks fell away, tumbling through the sky like spent organs. He craned his neck, saw them spiraling down and, for a moment, he whispered thanks. The quick maneuver undoubtedly saved his life, giving him just the impetus he needed to pick up airspeed. For when he glanced over his right shoulder, he saw the raging SAM streak past his six o'clock position, missing the F-16 by a razor-thin margin.

He was alive. But the sky was still on fire.

As Al-Tuwaitha lay burning below, Keith Rosenkranz's flight group headed home through a sky sharpened by steel and smoke. His job was done, but the 388th Fighter Wing's mission was far from over.

Meanwhile, back in the 401st Fighter Wing's slice of the sky, the Vipers kept coming in low and mean. Captain Mike Roberts had been dancing with danger for days. Flying with the 614th Tactical Fighter Squadron, he'd already cut his teeth on ground-attack sorties during the opening night of the war. By now, the hiss of SAM motors and the staccato rhythm of flak bursts were familiar. But nothing could brace him for what was waiting in the heart of Baghdad.

He was deep in the stack (part of the eight-ship formation led by Jeff Tice) throttling towards his designated target like a steel arrow. Below, the capital city crouched under a dome of radar nets and SAM sites. Still, Roberts felt like he'd threaded the needle. The sky ahead looked clean. No ominous blips or tones. Just the pulse of adrenaline and the hum of an engine that had carried him across a dozen horizons.

Then it hit. The cockpit screamed.

The unmistakable growl of a radar lock buzzed through his headset like a curse. A heartbeat later, his cockpit lit up with a stark, deadly warning: SAM launch. SA-6 incoming.

"Break!" he howled, slamming the stick into a gut-wrenching roll.

The Viper twisted hard. For a moment, he felt like he'd beaten it. The missile streaked overhead like a devil's spear, then detonated behind him, low and wide. "I thought it had missed," he later said.

But shrapnel from the blast had peppered the underbelly of his F-16. The engine coughed, then choked and died. The stricken jet lurched forward into a nose-down pitch, bleeding altitude fast. Alarms howled. Control was gone.

"I figured it was time for me to get out," said Roberts.

He jettisoned the canopy and pulled the ejection handle. A brutal blast of air tore him from the jet at 20,000 feet. Suddenly, the roar of air combat vanished, replaced by the surreal silence of freefall. As his parachute blossomed, the prevailing winds were kind, but he landed only a few hundred yards from a busy highway. Roberts grabbed his survival pack and ran towards what he thought was safety, but a furious crowd of armed civilians quickly surrounded him. "I put my hands up," he remembered, "and they kind of swarmed around me and took me."

They ripped through his gear, tearing away his helmet, his pack, and his comms. Only his flight suit and vest stayed on—not because the Iraqis were merciful, but because the garments were too complicated to strip off. Then came the soldiers. Iraqi troops arrived and shoved Roberts into the back of a truck, their eyes cold and expressionless. The interrogations would begin soon.

Jeff Tice, meanwhile, was about to have his own brutal encounter with an enemy SAM. His eight-ship Viper formation had knifed its way into the heart of Baghdad, deep into hostile airspace, under the reach of every missile battery Saddam's regime could throw at them. Their target: a massive oil refinery straddling the Tigris River. As Tice banked into his attack run, a strange stillness settled over the cockpit. It was the kind of calm that made his instincts twitch. "A proverbial sucker hole," he called it. A clear sky over a killing field. It felt like a trap.

Still, the Vipers unleashed their payloads, one after the other, bombs tumbling clean into the refinery's core. The spiring towers erupted in a chain of fireballs—"a beautiful series of secondary explosions," Tice remembered—mushrooming skyward with a black and orange medley as the fuel tanks cooked off below.

Tice let himself savor the victory for only a second. Then the sky lit up. Iraqi SAMs and anti-aircraft fire clawed their way from the ground, swarming the eight-ship Viper formation as they tried to egress. Tice yanked his F-16 into an evasive maneuver. But as he rolled out, his eye caught something no pilot ever wants to see: One of his squadron mates disappearing in a flash of flame.

It was Mike Roberts.

Tice had barely processed the event before the aircraft tore itself apart under the SAM's warhead, vaporizing midair. Fearing the worst, he didn't realize that Roberts had survived.

But it was no time to mourn. The sky was now a killing zone.

Missiles came screaming toward him. Iraqi radar sites were throwing everything they had. US Intelligence later confirmed that the Iraqis had launched between nine and twelve missiles at Tice's formation within the span of a single minute. His radar warning receiver was a chorus of panic—locks, launches, and proximity tones—each one a proverbial roll of the dice. He was pushing the F-16 to its evasive limits when one final missile closed in, fast and inescapable.

At the last possible second, he hauled the stick into a full-body barrel roll, snapping the F-16 under the incoming SAM. But at this proximity, it was no use. The warhead detonated just behind him. "I gave it a 32-foot day," he said afterward—his way of describing how close it had come to ripping him apart. Shrapnel hammered into the rear of his jet, "like an exploding cigar," he later recalled. The F-16's controls went limp, but the backup systems surprisingly kept it airborne. The bird was bleeding out, but not yet finished. Tice coaxed the ailing Viper southward, running on fumes, systems failing one right after the other. Every minute felt borrowed.

Finally, at 14,000 feet, the game was up. He pulled the ejection handle. In an instant, he was "face down in the cold air, canopy tumbling away, strapped in the ejection seat." The coalescing chaos of SAMs and anti-aircraft guns faded. Under his chute, he fell through the thin clouds, consciously aware that time was not on his side.

Under a drifting canopy at around 3,000 feet, Tice emerged from the clouds descending towards a nearby marshland. Not good. "In the Iraqi desert, water attracted people." He saw tents two miles off. What looked like campfire smoke turned out to be automatic weapons fire. Somebody was shooting at him. Tice hurriedly slipped his parachute into an evasive maneuver, a trick that most paratroopers use as a means to avoid ground hazards.

But these standard airman parachutes weren't exactly "maneuverable."

He hit the ground, cut away his chute, and started to run. But within a few "Jesse-Owens-style strides," his faceless attackers began shooting up the ground in front of him. He halted and threw his hands in the air. Roughly a dozen Bedouin tribesmen encircled him—not the elite Republican Guard or Iraqi Police, but the reclusive nomads who wandered the deserts of Arabia. "Most of them didn't have teeth or shoes," he said, "but they had brand-new AK-47s." The next day, they drove him into the nearest town and turned him over to the Iraqi Army, where he too was bundled away into captivity.

Within days of being captured, Jeff Tice and Mike Roberts were paraded in front of Iraqi state television cameras. They didn't need to say a word. Their eyes said enough. Beaten, bloodied, and visibly under duress, the two pilots stared out from the screen...faces drawn, hands clenched. But they were still alive. And that alone made the footage electric.

Their televised appearances aired from January 20-22, confirming to their families (and the US public) that both men had survived the shootdowns. But they were now prisoners of Saddam Hussein.

What came next was a descent into darkness.

Dragged into the brutal catacombs of Saddam's prison network, they were isolated, interrogated, and savagely beaten. By the end of the war, Tice and Roberts would be among the nearly two dozen American POWs taken captive by Iraqi forces.

Yet their experience became a beacon of hope. Back at the 614th Fighter Squadron operations center, their fellow pilots painted a sign reading "God Bless Tico and MR" (referring to Tice's and Roberts's callsigns, respectively) over the building's main entrance. Pilots leaving on missions would tap the slogan before takeoff. Indeed, their fallen brothers had ignited a fierce resolve. From that moment forward, until the end of Desert Storm, every sortie would become a tribute and a promise to their brothers in captivity.

On March 6, 1991, Jeff Tice and Mike Roberts were finally released by Iraqi authorities. After forty-seven days in captivity, both men were delivered to Allied forces during a silent handover, followed by a one-way flight out of the war zone. Reuniting with their families and squadron mates, the downed aviators were given a carrier bag filled with various treats, but their emaciated frames could barely digest them. Still, the moment felt like redemption: They were home and they were alive.

In time, both men returned to duty.

They walked back onto the flight line with shoulders squared and eyes forward, proud survivors of a harrowing chapter in the annals of combat aviation.

All told, the Package Q Airstrike had been a success, but not a resounding one. When the smoke cleared over

Baghdad, post-strike analyses revealed that many of the intended targets had been severely damaged or destroyed. Jeff Tice's bombs, for example, had hammered the designated refinery, knocking it offline in a rolling wave of secondary explosions. Elsewhere, the 388th's Vipers had zeroed in on the Al Tuwaitha complex, leaving it scorched and crippled, but not completely destroyed. Iraqi radar sites had been hammered, SAM batteries destroyed, and for a brief moment, even the heart of Baghdad had felt the full weight of American air power.

Yet the price hadn't been one-sided.

Two American F-16s had gone down. Two pilots—Jeff Tice and Mike Roberts—were dragged off at gunpoint by Saddam's forces. Their ejections had been violent, and their captures were even worse. Grainy footage of their bloodied faces flashed across CNN as the world got its first glimpse of American POWs. At home and on the flightline, the images had a profound impact.

For CENTAF, the message was clear. Baghdad wasn't just "dangerous;" it was a deathtrap for anyone flying without stealth technology. The city was ringed in iron: SAMs, radar-guided guns, and infrared seekers, all stacked like layers of a huntsman's trap. One postwar analysis put it bluntly: Package Q demonstrated that "the air defenses around Baghdad were dangerous for non-low-observable aircraft." In other words: Unless you were invisible, don't go back.

Thus, after January 19, the coalition shifted gears. Gone were the massive daytime raids. In their place came the stealth-driven, nocturnal attacks: Covert surgical strikes spearheaded by the F-117 and other ghosts of the night. Package Q had been a bold, brave, and ambitious undertaking. But it had come at a cost. In the end, some called it a tactical success, but a strategic wake-up call.

Inside the 614th Fighter Squadron, however, the mood was decidedly different. Tice and Roberts were down, but they weren't dead. Every heart in the squadron broke when Roberts' and Tice's jets vanished from the radar, but morale soared days later when both pilots re-emerged, alive, on international television. Captain Phil Ruhlman, one of their squadron mates, said it best: "In three days, we went from the lowest low to the highest high." Their survival became more than just a footnote. It became a banner, a rally cry for the remainder of the conflict.

Still, the legends lived on: Jeff Tice rolling through the SAM kill zone with a dead jet behind him; Emmett Tullia diving through six SAM launches with no working flares and no time to pray; and Keith Rosenkranz, cheating death while running the gauntlet of fire over Al Tuwaitha. These weren't just combat anecdotes. They were proof that even in the worst skies imaginable, a good pilot could still punch through and make it out alive.

Killer Scouts

By the end of January 1991, the air war had taken an interesting turn. The first week had been a relentless barrage. Iraqi airstrips had been cratered, hangars reduced to rubble, and hardened shelters set ablaze like funeral pyres. But now, CENTAF's intel briefings carried an unexpected twist: the Iraqi Air Force was vanishing. Radar tracks showed them bugging out—a sudden "mass exodus" of Iraqi aircraft fleeing to Iran. It wasn't surrender; it was self-preservation. Indeed, the Iraqi Air Force had been flying their jets across the border, hoping to use the shield of Iranian neutrality as a means to wait out the conflict.

With most of the Iraqi Air Force going MIA, the focus of the air campaign shifted to targeting the elite Republican Guard. Whereas the Iraqi Army subsisted on cannon-fodder conscripts, the Republican Guardsmen were purported to be Iraq's "best of the best." They were fiercely loyal to Saddam and had orders to fight to the last man standing. Their barracks were modern, their training was brutal, and their devotion was absolute. These priority killers were hand-picked, well-paid, and well-equipped with the best Soviet tanks and artillery that money could buy. If CENTAF wanted to affect the coalition's ground campaign into Iraq, they would have to destroy (or at least soften) the Republican Guard from the air.

But even from the third dimension, surgically dismantling the Republican Guard would be no easy feat. It wouldn't be as clear-cut as bombing an airfield or knocking out a mobile radar site. Rather, it would be a true battle of air power against armor.

Late in the afternoon on February 3, a returning sortie of F-16s from the 388th Fighter Wing landed at Al Minhad. They had just carried out another strike against the Republican Guard.

But something felt off.

The pilots climbed down from their cockpits in a sullen haze. Even the most swaggering jet jockeys among them had to admit the obvious: Today's mission had been unfulfilling. "They had missed targets," said Lieutenant Colonel Mark Welsh, "and felt certain they had mistakenly bombed some empty revetments." It hadn't been their first mission against the Republican Guard, and it hadn't been the first mission to feel like a bust.

Still, the pattern had become familiar. And the pilots

were getting frustrated.

For every sortie against the Republican Guard, the F-16s were running into the same set of problems. Every kill zone looked the same. Dozens of armored formations lay scattered across the desert, each one a carbon copy of the next. Tanks, artillery, and support vehicles were spaced in deceptive patterns. Decoys, camouflage, and mirage-like symmetry had become the rule of the day. Moreover, the sheer volume of coalition air traffic (and the lingering threat of enemy SAMs) meant there was no time for the F-16 pilots to loiter over the target area and assess the damage below. The missions had become mechanical: Target, Dive, Drop, Climb, Egress, Repeat…all while hoping the intelligence data had been correct.

But often, it wasn't.

"Not only were they unsure they had found the 'right' group of armored vehicles," Welsh continued, "but they also did not know if they had hit their targets."

Heavy cloud cover only made things worse. Entire strike packages would fly for hours, refuel from the KC-135s, then plunge through a wall of weather so dense that pilots could barely see their wingmen. By the time they broke through the clouds, they had little choice but to initiate their attack dives. A pilot could either drop his ordnance on what he *thought* was the right target, or he could abort the mission and return to base with his bombs unused.

Neither option felt good.

To this point, they had flown dangerous, low-level attack runs; braved venomous anti-aircraft fire; dodged SAMs; and evaded predatory radars. But what haunted them wasn't the risk; it was the ambiguity. It was the nagging suspicion that they had burned fuel, risked their lives, and blown up nothing but dirt, decoys, and empty targets.

Meanwhile, at CENTAF Headquarters in Riyadh, Brigadier General Buster Glosson and Lieutenant Colonel Clyde Phillips were addressing that very problem. Glosson had been appointed CENTAF's Director of Campaign Plans, and his immediate staff included a cadre of pilots from the USAF Fighter Weapons School, all of whom reported to Clyde Phillips. Coalition forces had been at war for seventeen days, and the latest intel confirmed that the "attrition of Republican Guard units was less than expected." Because the F-16s in theater were flying the highest number of sorties against the Republican Guard, Glosson began brainstorming ways to improve the Vipers' overall effectiveness.

To that end, Phillips took his cue from recent history. The problems confronting these F-16 pilots sounded oddly reminiscent of what their forebearers had seen in Vietnam. The confusion, the misidentification, the wasted ammunition—it all sounded familiar. As it turned out, American attack squadrons had previously solved this problem in 1966 by turning their high-performance aircraft into forward air controllers (FACs). These "Fast FACs," as they were called, flew ahead of the strike packages scanning the ground, designating targets, and calling out SAM threats. Fast FACs were aerial reconnaissance and attack coordination rolled into one lethal package.

Clyde Phillips leaned into that history like a lifeline. And he argued that, in the desert, the concept might work even *better*. He suggested that Glosson could use fighters to *locate* Republican Guard units, *call* in strikes, and then *assess* damage afterward. It would be a dynamic triangle of air power—hunt, coordinate, and confirm.

But the question remained: Which fighter would

assume the role? The mission demanded long hours deep in hostile airspace, defensive awareness, and tight navigational precision. The F-16 quickly rose to the top. It had the endurance, agility, radar, and multirole capability they needed. Glosson liked the idea, and he told Phillips to build the concept. But now came the bigger question: Which unit would spearhead the Fast FAC in combat?

The answer came quickly.
The 388th Fighter Wing had just sent Glosson a summary debrief of their lackluster raid from February 3. They weren't asking for more bombs; they were asking for an "airborne platform" to validate their designated targets, a secondary aircraft that could survey the battlefield *before* the main airstrike arrived. In essence: *They were asking for Fast FACs*. That message changed everything.

Glosson knew the 388th had the *Block 40* F-16Cs with GPS, the most accurate navbirds in theater. He asked if they had any experienced FAC pilots, to which the 4th Fighter Squadron happily identified sixteen of their pilots with previous FAC or A-10 support experience.

The blueprint fell into place: The 388th Fighter Wing would build the mission, while the 4th Fighter Squadron supplied the manpower. They showed Glosson their proposal which, serendipitously, was almost identical to the concept Phillips had drawn up.

The plan resonated with Glosson, but now they needed a name. Glosson didn't want to recycle the "Fast FAC" moniker; he was afraid it would cause confusion with A-10 FACs. He suggested the term "Scout," but Phillips made a clever observation. He pointed out the F-16s lacked rockets, and would have to mark their targets with 500-lb bombs. Thus, he proposed naming the F-16s "Killer Scouts." It wasn't elegant, but it sounded tough…and the name stuck. Late on February 3, CENTAF commander General Horner gave the green light. Training would begin the next morning over southern Iraq.

The mission profile was lean and lethal. "Killer Scouts would validate targets in the ATO that had been assigned to the F-16s and then find other lucrative targets in the area. They would provide indirect control, target area deconfliction, threat information, and updated target coordinates and descriptions to inbound fighters." Based on real-time information, Killer Scouts would either clear strike packages to hit the original target, or re-route them to a pre-designated backup target, all while coordinating with local AWACS and EC-130 Airborne Battlefield Command & Control Centers. Every inbound fighter would get on the Killer Scouts' frequency, and every strike pilot would fly under their watchful eyes.

Target areas were selected according to the ATO; and each target area was annotated within a pre-existing CENTAF "kill box." Every kill box had an area of 900-square nautical miles, each represented by grid overlays on a map of Kuwait and southern Iraq. These grids brought simplicity to the chaos of aerial warfare: Fighters knew exactly where to go, and planners could focus their firepower where it mattered most.

The 388th selected eight pilots for the inaugural Killer Scout missions. They went out with charts, target lists, and fear in their hearts; for they would now be orbiting Republican Guard territory for hours on end. They adopted the call sign "Pointer." Their job: to *point* incoming strike fighters towards the best targets.

At sunrise, February 4, the first two-ship Pointer team hit their kill box. They stayed eight hours in the air, rotating between their kill boxes and a designated tanker track in the Persian Gulf. And the F-16 pilots were stunned by what they found: Iraqi combat vehicles and other ground weapons lay buried under the sand, hidden within cleverly-dug revetments. In fact, some of the Iraqi units identified in that day's ATO had already relocated. They were long gone, nothing left but empty bunkers. Undeterred, the Killer Scouts simply found *new* targets: Assembly areas, ammo dumps, and artillery pits.

The F-16s logged them all.

Then, the follow-on strike package hit. In a two-hour blitz, 120 coalition aircraft screamed into the box under guidance from the Killer Scouts. Between strike surges, the Pointer F-16s circled back to the blown-up targets, assessed the damage, and recommended follow-up attacks if needed. For the first time since January 17, real-time battle damage assessment was being done on scale.

From there, the momentum grew.
On February 5-6, Team Pointer repeated the pattern. Glosson's team, meanwhile, refined the comms plan, kill-box layouts, and fighter flows. Pointer flights expanded to cover six kill boxes concurrently, covering some 5,400 square nautical miles of desert. What started with eight pilots became thirty-two sorties per day. And almost every pilot in the 4th Fighter Squadron would run a mission under a flight leader fully-qualified in the Killer Scout tactics.

Their missions were long (five-and-a-half hours) stacked with three, one-hour target blocks and four aerial tankers. A flight lead and his wingman would fly in a "shooter-cover" formation: high-altitude scans, steep dives, bomb drops, and secure recovery flights. One F-16 would roll in low to light the target; the other stayed high, watching his radar for any SAM threats. The Killer Scout's bombs could then mark, smash, or suppress the target.

If a SAM battery lit up in their zone, it would become the first priority kill. Major JD Collins, another F-16 Killer Scout, demonstrated the typical rapid-response, counter-SAM tactics during an early February mission. That day, he had observed an Iraqi SAM launcher firing an SA-2 at an egressing B-52 over his area. Luckily, the SA-2 missed the heavy bomber by a wide margin.

But Collins wasn't going to let this SAM crew escape the wrath of his F-16. Within moments, he reached the SAM site, whereupon he saw a "single launcher being pushed back into a large tin shed and a radar van parked nearby." He and his wingman happily pounded the site, then directed the next three strike packages to hit it again...just to drive the point home.

It worked: Daytime SAM launches nearly vanished from the theater. And if a Killer Scout couldn't suppress a SAM site or anti-aircraft battery on its own, escort jets and F-4G Wild Weasel flights would be summoned for the task. Scout planners would brief the F-4Gs, then circle to the next grid cell, keeping the pressure on Iraqi ground forces everywhere.

But, at the same time, target camouflage became common. Decoys multiplied and dunes were carefully crafted to hide revetments. Sun angles often turned shadows into mystery. Under these conditions, dropping in on a kill box meant binocular-level scrutiny. Still, these Killer Scouts hung on every detail. When weather pushed them above lower clouds, they switched to "Ground-Moving Target" radar mode. This allowed the F-16s to drop radar-guided bombs from medium altitudes with pinpoint precision.

When the ground offensive began in late February,

Allied maneuver speeds skyrocketed. Indeed, some ground units were travelling in excess of twenty miles per hour—an impressive speed for armored and mechanized units. However, this meant the Killer Scout grids needed real-time coordination with A-10 FACs, Army units, and various AWACS. One wrong pass, one misread target, and a strike jet could bomb a friendly armored column. Killer Scouts became that link, connecting ground reality with multi-layered air power.

The impact was palpable. The Republican Guardsmen didn't dare move by day. And whenever they moved by night, the sundown bombers and Special Operations aircraft inevitably found them. By the end of the war, Glosson remarked that the Killer Scouts had increased the F-16's effectiveness "three- or four-fold." Today, every American F-16 squadron has incorporated the Killer Scout methodology into their training program, complete with six certified lead pilots. From the throes of Desert Storm, the 388th Fighter Wing had taken the F-16 from fighter to hunter. And in doing so, they had changed the game forever.

"Killer Scouts also improved feedback to CENTAF headquarters," said Bill Andrews, the 10th Fighter Squadron pilot who had been attached the 363rd Fighter Wing. Andrews recalled General Glosson's communique from February 14, 1991, confirming that the F-16 Killer Scouts had more than doubled the number of Viper-reported kills per daily mission cycle. In other words, the number of enemy ground losses attributed to F-16 strike operations had *doubled* within a mere ten days.

Still, their success rate was never 100%.

"One critical problem Killer Scouts did not normally resolve," Andrews later admitted, "was that of precise target discrimination"—largely attributed to the Republican Guard's deception efforts. Although Killer Scouts were often the pilots most familiar with the area surrounding Northern Kuwait and Southern Iraq, "there were limits to what could be discerned," said Andrews, "and some Iraqi deception measures [e.g. decoys] were very difficult to penetrate."

While most of the Killer Scout missions were a resounding success, others ended in disaster. Bill Andrews would experience both. On February 24, 1991, he was flying lead in a four-ship formation on what should have been a "fairly routine mission," as he called it. "We were supposed to hit armored formations," but during his mid-air refuel, Andrews and his wingmen received new coordinates from another Pointer crew farther north. Punching the data into his navigation system, however, Andrews was shocked to see that his new target was more than *twenty minutes* due north...far beyond where any armored forces were templated to be.

Acknowledging the call, Andrews switched over to the emergency radio frequency. "It sounded like Pointer was talking to an Army unit on the ground and we were flying into an emergency CAS [Close Air Support] situation. Some people are in a bind...and we were going to help them out." Little did he know that a Special Forces A-team had been surrounded and were now fighting for their lives. What should have been a routine airstrike had transformed into a desperate rescue mission deep in Republican Guard territory.

Through the radio chatter, Andrews and his wingmen discovered that Special Forces Operational Detachment Alpha-525 (eight Green Berets led by Chief Warrant

Officer Richard Balwanz) had been discovered by local civilians. Now, enemy forces were converging onto the beleaguered Green Berets as they stood their ground less than 200 yards from enemy lines. Pinned down and running low on ammunition, the A-team had launched an emergency UHF GUARD radio distress call, their voices brittle with fear and determination...and Andrews's flight was en route to deliver them from the storm.

Smoke billows and swirling sands were casting a grim haze over the target zone. Visibility was poor, and determining friend from foe would be nearly impossible.

Yet, Andrews didn't hesitate. He flew headlong into the lion's den.

He rolled in, cluster bombs rattling in the racks. The first volley unleashed a storm of bomblets among the enemy formations. Dust roared upward, and the enemy's response staggered. But the situation demanded more.

Andrews circled and dove again.

Enemy fire opened up. Still, Andrews pushed the attack. As the F-16 came in for its second dive, one of the Green Berets, Sergeant First Class Robert DeGroff, looked skyward and yelled to his comrades:

"Take cover! Take cover! This is going to be close!"

A little too close, it seemed.

Indeed, Andrews's angle of attack had the perfect markings of a "Danger Close" mission—wherein live ordnance (bombs, artillery, naval gunfire, etc.) are dropped to within 600 meters of friendly troops in contact. These Danger Close fire missions were intended as a last resort for troops who were in danger of being overrun.

And today's mission certainly warranted the procedure.

"We jumped down in the ditch and pressed against the

side of the hard dirt wall," DeGroff continued. "I peeked over the rim and watched as the jets came down in slow motion, just like in a movie. They dropped and the bombs broke open and the pellets spread out just like they were coming right at us. It was awesome. The earth turned to Jello as the bombs went off—BOOM! BOOM! BOOM! The earth just shook while the pellets kept exploding right outside our ditch."

Meanwhile, the enemy formations faltered. Iraqi troops, daringly bold just moments before, were now scattering for cover under Andrews's relentless assault. Smoke pillars, dust, and tumbling metal confirmed the suppression, but Andrews stayed overhead long enough to ensure the pressure and shock effect remained. Suddenly, the rescue helicopter thrashed into view—it, too, drawing enemy fire—but the worst of the Iraqis' firepower had been broken. The ODA scrambled aboard, their lives pulled back from the jaws of death.

Andrews loitered over the battlespace until the chopper disappeared over the horizon. Only then did he disengage and exit the smoke-filled kill box. For his actions that afternoon, Andrews received his second Distinguished Flying Cross with "V" Device for valor, denoting his remarkable courage in saving American lives under extreme duress

DeGroff later commented that the F-16s gave him and his A-team a calm sense of control. "It was like watching them on a TV screen where you know how the show ends." He admired the deft airmanship and the drill-like precision of their targeting runs. "If I told them to bring it in 100 meters, that's exactly where the bombs hit." And today, within just a few heart-pounding moments, Andrews delivered his ordnance within a jaw-dropping

200 meters of the friendly position—well below the doctrinal threshold of 600 meters, but absolutely necessary to keep friendlies alive.

Nothing about the mission had been planned. It was all adaptation: A fighter pilot turned emergency FAC, guiding rescue efforts, bombing kill zones with surgical control. It was the gold-standard for the F-16 as a hunter-killer FAC—a fusion of reconnaissance, command, and strike capability all within a single airframe. That mission of February 24 would be studied in flight schools and cited in manuals for years to come. It embodied what frontline air power *could* be when men were willing to bend rules, fly through fear, and coordinate complex operations under fire.

For Andrews, the mission gave him pause to reflect. He later asked himself:

"Well, I know for sure I killed a lot of the enemy. How do I feel about it?"

To which he answered: "It was them or us."

In the unforgiving game of war, Andrews knew that he would have to kill the enemy before the enemy killed him, or his comrades. "That sort of resolved for me any moral ambiguity over this mission and the fact that we were killing the enemy," he said. "It was sort of a watershed in my attitude towards the war, and the enemy, because pilots have the ability—usually the luxury—to consider the job…and think about things in terms of targets instead of people." According to Andrews, bombing a building from 20,000 feet was different than flying a Danger Close mission a few hundred feet from the ground, with real-time visual contact of the people he was attacking.

But such was the nature of war. At the trigger-pulling level, it truly was a matter of: "It's us or them."

In the final week of February 1991, Desert Storm was reaching its crescendo. The long-awaited ground campaign began on February 24, spearheaded by the US VII Corps. By February 27, Iraqi forces were collapsing. Armored columns broke, artillery units fled, and the Republican Guard was preparing to make its last stand. Meanwhile, Bill Andrews was back in the sky, leading his F-16s on a hunt for retreating enemy armor.

On his way back to base, however, the desert finally bit him back.

The radar flickered. SAM launch indicator.
Andrews banked hard, looping a hard G-turn just as an SA-2 growled into the air. The missile streaked past, then arced beneath him before detonating, pushing him against the canopy as he pulled staggering G-forces. The engine gurgled; and the master alarms blared.

And he ejected, violently.
Below him lie the savage, sprawling terrain of the Iraqi badlands. Mere seconds after feeling the jolt of his fully-bloomed parachute, the desert floor hit him with a vengeance, breaking his right leg. He had landed deep inside Republican Guard territory...and the odds were not in his favor. Suddenly, voices rumbled in the distance, and ground fire began crackling around him. The Republican Guard had found their man.

He couldn't stand; he couldn't run. He was alone, broken, and behind enemy lines. He tried to raise his arms as a group of Iraqi soldiers stalked toward him with their AK-47s. They barked commands in Arabic, motioning him to get up. But all he could do was gesture towards his shattered leg. The Iraqis crept forward cautiously, their AKs solidly trained on his battered body.

That's when he saw it: The unmistakable puff of smoke

and the glowing line of an SA-2 rising into the sky. An Iraqi SAM battery had just fired on another F-16 orbiting overhead. Andrews could see that it was his wingman trying to save him. But Andrews was flat on his back, his F-16 in tatters, nursing a broken leg...and he didn't want his wingman to join him.

"I'm in a world of hurt; I don't want any company; I've got to do something," he thought. Realizing he had only seconds to react, Andrews hastily reached for the survival radio tucked against his gear.

"Break right, flare, flare, flare!" he shouted into his handset.

The F-16 above snapped into a break turn, flares tumbling in rapid succession. It was a gutsy move on Andrews's part, considering the Iraqis had nearly thirty automatic weapons trained onto him. But he couldn't let his wingman get shot down, too.

The Iraqis, however, voiced their displeasure with gunfire. Bullets tore up the dirt around Andrews. He threw the radio down and tried again to raise his arms. "I think I shouted, 'They're attacking me!'" he later recalled. The riled Iraqis swarmed him, stomped his radio, riddled his helmet, shredded his gear, and manhandled him into the back of a waiting jeep. He was now a prize of war, marked for delivery to Basra.

But his transport didn't make it far.
On a dusty road, the jeep's engine kept stalling. Then it died again. And again. When it stalled for the tenth time, the driver stepped out to check under the hood. That's when Andrews saw an imperfect gift delivered at the perfect time: Glinting copper and gold, a wave of bomblets came raining down from the sky. An American F-16 had just dropped a CBU-87 cluster bomb right where their jeep

was supposed to be.

Timing had saved him.

The stalling engine, ironically, had spared their lives. The driver scrambled back inside, eyes frozen wide in panic, and floored the pedal off-road into the desert. That night, the Iraqis dragged him into a dirt bunker and splinted his leg with bamboo.

At some point during the night, he drifted off to sleep, before being startled awake by the sound of Iraqis yelling and running around. They were packing up, clearly in a panic. They hauled Andrews outside and dropped him next to a vehicle, but didn't leave a guard. In fact, they didn't seem interested in watching him anymore. Moments passed and the Iraqis seemed to forget he was there. Weighed down by a busted leg and 200 pounds of combined body weight and personal gear, Andrews began crawling nonchalantly...inch by inch...dragging himself back towards the bunker. As soon as he reached the bunker's edge, he tucked himself underneath a discarded tarpaulin, and held his breath.

The Iraqis never saw him slip away.

Then came the rumbling of engines, which slowly faded into the distance. The Iraqis had driven off, unaware or unconcerned that they had just misplaced an American POW. For a moment, Andrews thought he had made it. But then, he thought: *"OK, genius, you've got a broken leg in the middle of the desert. What now?"* He fashioned a white flag from some cloth and hoped for a rescue. All throughout the night, he heard the sound of artillery booming overhead—Allied forces were getting closer.

But freedom was fleeting.

The next morning, Iraqi patrols swept the area, probably looking for deserters. One patrol lifted the canvas flap under which he was hiding, and promptly dragged him out. This time, they made it to Basra. At a local hospital, Iraqi medics realigned his leg. Then, they blindfolded him and tossed him onto a bus headed for Baghdad.

Through the crevices of his blindfold, he could see traces of American uniforms. He was not alone. Although he was ordered not to talk, he mustered the courage to make a request. When a cold wind blew through an open window, Andrews asked aloud: "Captain William Andrews with a request, sir. The wind is cold. Will you close the window, please?" The nearest Iraqi guard honored his request with a reluctant grunt, but Andrews had been thinking strategically: He had said his name out loud, hoping it might resonate with another Allied prisoner on the bus.

Serendipitously, it did.

A voice from behind whispered: "Airborne."

Andrews later learned he was seated just ahead of Sergeant Troy Dunlap, a US Army pathfinder from the 101st Airborne Division—part of the search-and-rescue team that had been diverted to find him. Their helicopter had been shot down, and five of the crew had perished. "It's one of the hardest things, knowing that others died trying to save you," Andrews said. Among the survivors of the ill-fated crew, however, was Major Rhonda Cornum, an Army flight surgeon. Coincidentally, her husband, Kory Cornum, was an Air Force flight surgeon attached to the 58th Fighter Squadron, an F-15 Eagle unit forward-stationed in Tabuk, Saudi Arabia. The 58th would go on to make headlines as the top-scoring American fighter squadron of Desert Storm—shooting down more Iraqi MiGs than any other unit.

Arriving at an intelligence facility in Baghdad, the real

horror began. The Iraqis tried to soften him up. "Look, we know you couldn't talk in front of the other Americans. Now you can talk. Nobody will ever know."

He refused. That's when the beatings started.

Two interrogators worked him over, emotionally and physically. It was a war within a war—one of endurance, principle, and pain. "You get through one session and think you're OK," Andrews recalled. "Then another one starts. And you think: What can I even tell them? That I flew an F-16? They already knew that."

Then, the war ended.

On February 28, President Bush announced a cease-fire to the ground war. Barely 100 hours after the start of the Allied invasion, the Iraqis were in full retreat and Saddam was desperate to sue for peace. His much-anticipated "Mother of All Battles" had come to pass—but it was the Iraqi military that had been routed. In their disastrous retreat, the Iraqis had fled Kuwait, igniting oil fields in a vain attempt to slow down advancing coalition forces. Come what may, it would take a massive reconstruction effort to get the emirate back on its feet; but for now, the savagery of Iraq's occupation had ended.

On March 3, 1991, General H. Norman Schwarzkopf, commander of UN Forces, met with several Iraqi generals in Safwan to discuss the terms of surrender. During that meeting, Schwarzkopf made it clear: POWs would be returned in good condition, or there'd be hell to pay.

The Iraqis got the message.

Bill Andrews was released alongside Jeff Tice, Mike Roberts, and several others. As their Red Cross flight took off, the cabin was silent. But when two F-15s pulled up on either side, escorting them out of Iraqi airspace, the former

POWs erupted with cheers.

For his courage under fire, and for protecting a fellow pilot while wounded and captured, Bill Andrews was awarded the Air Force Cross. He had gone from hunter to hunted, from pilot to prisoner. But even on his back, and surrounded by enemies, he had retained his fighting spirit. Like Tice and Roberts, Bill Andrews returned to flight duty—a proud survivor and quiet witness to the war's darkest hour.

F-16 "Wild Weasels" & F-4G Phantoms

By the time Desert Storm began in January 1991, the US Air Force had perfected a deadly partnership: The aging but still formidable F-4G Phantom II and the agile, modern F-16. Together, they formed the backbone of America's "Wild Weasel" force—specialized units combining *Suppression of Enemy Air Defenses* (SEAD) with *Destruction of Enemy Air Defenses* (DEAD) by way of anti-radiation missiles. This unlikely partnership remains one of the lesser-known, yet most-critical missions performed by F-16s during the Gulf War.

The role of the Wild Weasel was brutally simple. Fly into enemy airspace, bait SAM systems into turning on their radars, then kill them before they could launch. It was a deadly game of cat and mouse, and the consequences of failure were swift and unforgiving. For years, the Wild Weasel mission had been the domain of the F-105 and F-4, but by the late 1980s, the Air Force began pairing the two-seat F-4G with the newer and more maneuverable F-16C.

The F-4G, known as the "Advanced Wild Weasel," had replaced its gun with an extensive suite of radar-homing equipment. In the rear cockpit sat an Electronic Warfare

Officer (EWO), operating sensors that could detect and identify enemy radar emissions in real time. The aircraft could launch AGM-88 HARMs, the coveted High-Speed Anti-Radiation Missiles designed to home in on those radar signals and destroy the source. But the Phantom was slow—and by modern standards, quite heavy.

Enter the F-16C *Block 30/32*. While not specifically designed for SEAD missions, the F-16 was nonetheless adaptable. Communicating with the F-4G, the Viper could carry the HARMs itself, adding speed, agility, and enhanced survivability to the mission. Under these parameters, the F-4G could sniff out the threats, while the F-16 could engage, flank, or finish the job.

In Desert Storm, this tag-team approach proved to be a winning formula. F-4G/F-16 Weasel teams flew hundreds of sorties into the most heavily defended parts of Iraq—Baghdad, Basra, and key strategic installations—drawing fire from SA-2, SA-3, and SA-6 sites and eliminating them one by one. The F-4Gs led the hunt, locking onto enemy radar signals, while the F-16s flew close escort, ready to deliver HARMs or suppress anti-aircraft threats.

This coordination demanded precision and trust. Pilots often flew in mixed two-ship or four-ship flights, with F-4Gs and F-16s in constant communication. Timing was everything: Too slow, and a SAM might launch; too fast, and a radar might shut down before the HARMs could home in. And the risk was immense. These crews deliberately exposed themselves to the deadliest part of the enemy's integrated air defense system. But the payoff was clear: once the SAMs went dark, follow-on strike packages could reach their targets with far fewer losses.

Among the intrepid Wild Weasels was Captain Dan

Hampton—a young F-16 pilot assigned to the 23rd Fighter Squadron out of Spangdahlem Air Base, Germany. As Hampton recalled, the 23rd was the "only mixed F-4G/F-16C squadron in the world." Their parent unit, the 52d Tactical Fighter Wing was a single-mission, all-missile combat unit "trained to fly in the low-altitude European war that never was," he continued. "Yet, we were sent to a different theater...to go to war against a vastly different type of enemy." Indeed, he found himself based at Incirlik, Turkey, conducting attacks from the north against Iraqi SAM sites, radar stations, and anti-aircraft guns.

On February 19, 1991, Hampton flew one of his most harrowing missions of the war: A broad daylight raid into downtown Baghdad, designed to strike Command & Control targets. Previously, CENTAF had suspended daylight strike operations over Baghdad for conventional, non-stealth aircraft following the Package Q Airstrike. The F-117 Nighthawks had thereby reassumed the mission for attacking downtown Baghdad, but only under the cover of darkness. By early February, however, the Coalition resumed a limited set of daylight operations over Baghdad for specialized missions, particularly Wild Weasel flights.

The decision to reinitiate daylight airstrikes had been based on evolving intelligence, tactical necessity, and technological confidence. For one, Iraq's air defense grid was showing signs of recovery. After several weeks of pounding, some mobile SAM systems had begun repositioning and reactivating around the Baghdad area, particularly SA-2s and SA-3s. Their reappearance during daylight hours became a concern. Coalition leadership wanted to proactively target and suppress these reemerging threats to maintain air superiority—especially as ground operations were approaching. With ground combat preparations underway, CENTCOM needed more daytime aerial reconnaissance and more airstrikes to soften Iraqi ground forces. Better daylight visibility was sometimes necessary to identify mobile targets. Thus, to clear the way for daylight interdiction strikes near the Kuwaiti theater of operations, Baghdad-based SAM threats needed to be neutralized—especially in the daylight. Moreover, CENTAF was confident that F-4G/F-16 Wild Weasel teams could survive the daylight incursions by way of their APR-47 radars and AGM-88 HARM missiles. Lastly, CENTAF saw these operations as a controlled risk. Unlike Package Q (which involved dozens of aircraft moving in formation) a daylight Wild Weasel strike would involve a handful of F-4Gs, F-16s, and support jammers, moving fast, flying low, and hunting emitters. The risk was considered acceptable, especially since the crews were trained for exactly this kind of work.

Still, Hampton would describe the mission as "one of the hairiest" of his career. Although it was a "controlled risk," the Wild Weasel game was part of a high-stakes strategy to hammer Saddam's capital with precision weapons while luring out any remaining SAM launchers.

Flying as part of a composite SEAD package, Hampton operated in tandem with the venerable F-4Gs—all of which served as the primary EW platforms, while the F-16s flew close escort and carried AGM-88s to destroy the radar emitters once located. As they approached Baghdad, Hampton described how his radar lit up with multiple threat spikes. He described it as a moment of sensory overload: SAM signals flickering across his display, heat signatures spiking, and the ominous growl of missile warning tones. Iraqi SAMs were hurtling into the sky like Hell's own fireworks, tearing through the smoke with a vengeance that felt almost personal.

At one point, Hampton described the missiles cutting upward through the sky like "telephone poles," with one passing so close it left a vapor trail across his canopy. Startled, but not shaken, he executed hard evasive maneuvers while simultaneously firing his HARMs at radar sites identified by the F-4Gs and his own onboard HARM Targeting System (HTS). According to his account, several SAM radars were destroyed or damaged during the mission.

Like many of his comrades on missions before and after, Hampton described the flak over Baghdad that day as "thicker than soup"—noting that the sky had turned from blue to steel gray with smoke and explosions. His F-16 was rocked by near-misses as he dove and rolled through the anti-aircraft fire, maintaining speed and altitude just long enough to launch his HARMs before jinking away to safety.

Though the F-16 was originally designed as a lightweight multi-role fighter, Hampton and his squadron proved it could serve effectively in the SEAD role when paired with a platform like the F-4G. The two aircraft were rarely separated during missions. F-4G back-seaters (EWOs) would track enemy radars, often forcing Iraqi SAM crews to blink their systems off and on to avoid detection. The F-16s, with their speed and agility, however, could move in for the kill once those emitters were located.

The cooperation was seamless. Hampton flew dozens of combat sorties during Desert Storm, often deep into enemy territory, but the February 19 raid stood out as one of his defining moments. He later wrote that, after that mission, his flight suit was soaked through with sweat, his limbs exhausted from the sustained high-G turns and missile evasions.

Despite the dangers, Hampton's jet returned intact, and his HARMs had struck the enemy radars as planned—contributing to the broader campaign of destroying Saddam Hussein's integrated air defense network. Hampton's mission highlighted both the vulnerability and necessity of flying into hostile territory uninvited, as well as the unique role of the Wild Weasels in clearing a path for strike aircraft to do their work.

This was also a transitional moment in air power history. The F-4G was nearing retirement, and Desert Storm marked its last major combat deployment. The F-16, by contrast, was just coming into its own as a SEAD platform. Hampton's unit, the 23rd Fighter Squadron had trained relentlessly under the F-4G/F-16 fusion model. That training and coordination had paid off in full.

By the end of Desert Storm, the F-4G/F-16CJ teams had flown more than 2,000 SEAD sorties, helping neutralize more than 250 SAM sites and countless radar stations. Hampton flew thirty combat missions during the Gulf War, earning a Distinguished Flying Cross for his role in the Wild Weasel campaign. By war's end, the F-4G/F-16 Wild Weasel teams had helped neutralize much of Iraq's air defense network. It was the last combat stand of the mighty Phantom, but also the beginning of the F-16's rise as the Air Force's premier SEAD platform.

When the dust of Desert Storm finally settled in the spring of 1991, the F-16 had once again carved its name into the annals of air combat history—not in dogfights against enemy MiGs, but by excelling in the unforgiving realm of ground attack. Though the Viper never locked gunsights on an enemy plane in the skies above Iraq, it proved its worth day after day. They strafed, bombed, and scouted deep into hostile territory. On every mission, through every dust storm and SAM gauntlet, the Fighting Falcon delivered precision and persistence, reminding the world that air superiority isn't won just by aerial victories, but by relentless dominance of the battlefield below.

The pilots who flew the F-16 during Operation Desert Storm were a unique breed. Many were young, some barely in their twenties. They flew into the jaws of death at 500 knots, eyeballing targets across barren wastelands and cities bristling with SAM rings. For them, the cockpit wasn't just a workplace. It was a crucible where skill, technology, and nerves were tested under fire.

They had flown deep into the heart of Iraq, hammering airfields, radar sites, bridges, and hardened shelters. And when Saddam's air defenses sprang to life, the Viper pilots didn't blink. They simply adjusted, improvised, and struck again. Pilots like Jeff Tice, Mike Roberts, and Bill Andrews (all of whom were shot down and captured) carried the scars of war not just on their bodies, but in their souls. Yet even in captivity, they represented the courage and professionalism that defined the F-16 community.

In the Package Q airstrike, it was the F-16 that shouldered the burden, dropping thousands of pounds of ordnance on Baghdad while navigating a gauntlet of SAM and anti-aircraft fire so thick it turned the sky black. It was, as one pilot said: "The scariest rollercoaster ride in the world, except someone's shooting at you the whole time."

But Desert Storm wasn't just about raw firepower. It was about evolution. The war revealed shortcomings in aerial tactics and survivability—sparking upgrades to the targeting pods, precision weapons, and electronic countermeasures that would shape the Viper's future. It was in these hotly-contested skies that the F-16 earned its second reputation—not just as a bomb-dropper, but as a precision strike aircraft, a hunter of Scuds, and a true multirole platform.

When the last jets returned home in 1991, one thing was clear: The F-16 had set a new standard for American air power. It wasn't the newest or fastest fighter jet, but it had proven itself indispensable. Now, the Viper was more than just a fighter; it had become a symbol of American resolve and tactical flexibility.

Desert Storm may have ended, but the legend of the F-16 had only just begun.

Captain Mike Roberts sits in the cockpit of his F-16 from the 614th Tactical Fighter Squadron. Roberts was one of eight US Air Force pilots shot down and captured by Iraqi forces during Operation Desert Storm. He endured six weeks of interrogations and torture before being released. *US Air Force*

Photo Section

The prototype YF-16 (front) and YF-17 (rear). Both designs were contenders for the US Air Force's Lightweight Fighter Program (LWF) in the 1970s. General Dynamics' YF-16 was the winner; Northrup's YF-17 eventually became the F/A-18 Hornet. *US Air Force*

An F-16 from the 421st Fighter Squadron, Hill AFB, Utah, fires an AIM-120 Advanced Medium Range Air-to-Air Missile (AMRAAM) over a training range, 1994. *Rob Schleiffert*

F-16A *Netz 107* of the Israeli Air Force. *Netz 107* holds the record for the highest number of kills attributed to a single F-16. It bombed the Osirak nuclear reactor in 1981, and shot down six and a half enemy fighter jets (six solo kills, one joint kill shared with another Israeli F-16). Today, Netz 107 stands on display at the Israeli Air Force Museum. *Zachi Evenor*

Close-up of the kill markings on the fuselage of *Netz 107*. Five full roundels of the Syrian Arab Air Force denote the solo air-to-air kills, while the half roundel denotes the shared aerial victory. The triangular symbol depicts the 1981 attack on the Osirak nuclear reactor. *Oren Rozen*

Israeli Air Force pilots (from left to right, foreground to background): Hagai Katz, Ilan Ramon, Relik Shafir, and Amos Yadlin in the weeks leading up to Operation *Opera*, 1981. Opera was the daring daylight airstrike against Saddam Hussein's Osirak reactor. *Israeli Defense Force*

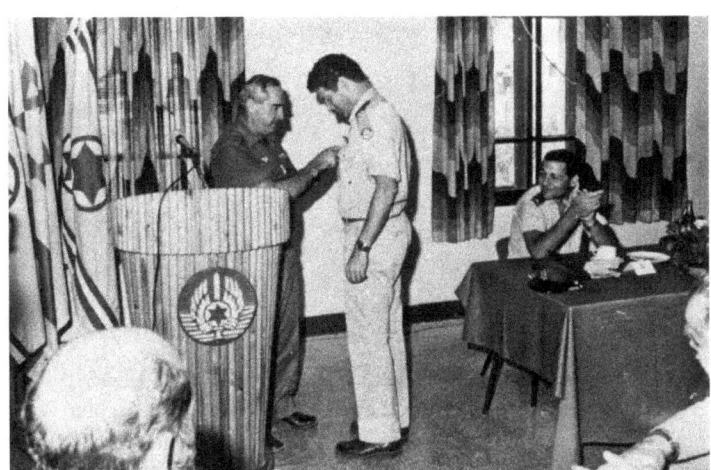

Israeli pilot Zeev Raz receives the Medal of Valor on June 16, 1981, for his role in leading the F-16 task force during Operation *Opera* the week prior. *Israeli Defense Force*

An American F-16C from the 388th Fighter Wing, prepares for a strike against targets in Iraq and Kuwait during Operation Desert Storm, 1991. *US Air Force*

A ground crewman guides another F-16 from the 388th Fighter Wing onto the taxiway. This photo was taken during the initial deployment phase to Saudi Arabia as part of Operation Desert Shield. This F-16 is equipped with an AN/ALQ-131 Electronic Countermeasure (ECM) pod and a Low Altitude Navigation, Targeting Infrared Night (LANTIRN) navigation pod. *US Air Force*

Ground personnel prepare to load Mk-84 2,000-pound bombs onto the F-16s from the 401st Fighter Wing. These aircraft were being armed for the Package Q Airstrike, the first daylight raid of Desert Storm, January 1991. *US Air Force.*

Pilots of the 614th Fighter Squadron listen intently during a pre-flight briefing before the Package Q Airstrike, January 1991. *US Air Force*

Jeffrey Tice, an F-16 pilot from the 614th Tactical Fighter Squadron. Tice was shot down alongside fellow pilot Mike Roberts during the Package Q Airstrike on January 19, 1991. Tice and Roberts were among the eight US Air Force pilots shot down and captured by Iraqi forces during Desert Storm. *US Department of Defense*

The wreckage of Jeffrey Tice's F-16. Tice safely ejected from the aircraft, whereupon he was captured by Iraqi forces. He and the other American POWs were later repatriated to the US after the war. The remains of his wreckage, however, were not recovered until the start of Operation Iraqi Freedom in 2003. *US Air Force*

Patrolling Iraq's No-Fly Zones. An American F-16J refuels near the Saudi border during Operation Southern Watch, 1998. *US Air Force*

Lieutenant Colonel Gary North (center), flanked by his crew chief, Staff Sergeant Roy Murray (right), and assistant crew chief, Senior Airman Steven Ely. All three men are posing with the F-16D that North flew when he shot down an Iraqi MiG over the Southern No-Fly Zone on December 27, 1992. *US Department of Defense*

A Royal Netherlands Air Force F-16 taxis out for a mission in support of Operation Deny Flight, enforcing the UN-mandated No-Fly Zone over Bosnia-Herzegovina, 1994. *US Department of Defense*

Captain Scott O'Grady (center) speaks at a press conference. While flying in support of Operation Deny Flight, O'Grady's F-16 was shot down over Bosnia on June 2, 1995 by Serbian forces. For six days, he evaded Bosnian Serb troops before being rescued by heliborne Marines. Captain TO Hanford (left) made the initial radio contact with O'Grady before the rescue. Captain Bob Wright (right) was O'Grady's wingman the day he was shot down. *US Department of Defense*

Peter Tankink of the Royal Netherlands Air Force. As an F-16 pilot, he scored the first aerial victory for the Netherlands since World War II. On March 24, 1999, during the first night of NATO's Operation Allied Force over Kosovo, he shot down a Serbian MiG-29 with an AIM-120 AMRAAM missile. *Netherlands Ministry of Defense*

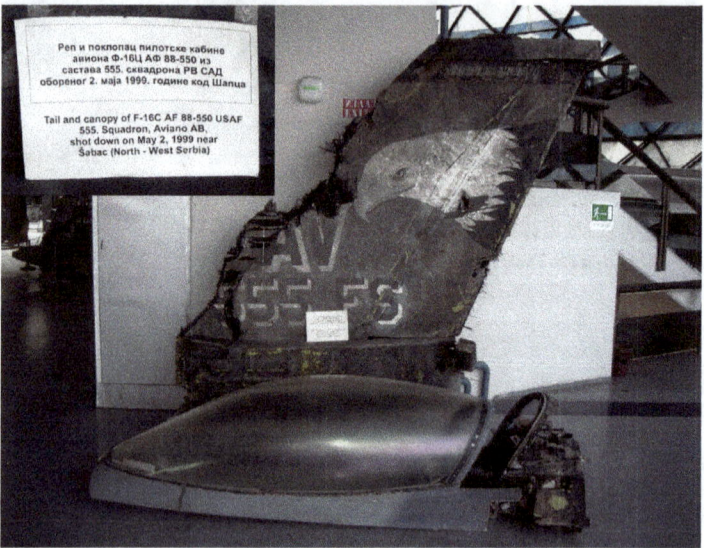

Tail and canopy of the F-16 piloted by David Goldfein, on display at the Museum of Aviation in Belgrade. Goldfein was shot down on May 2, 1999 during Allied Force, but successfully ejected. *Marko RFI*

Elaborate tailfin art of the Venezuelan F-16. First acquired in 1983, the Venezuelan Air Force used the F-16 to great effect during the attempted revolution of 1992. In more recent years, it has also served in a drug enforcement capacity, intercepting *traficante* flights that violate domestic airspace. However, a long-standing arms embargo has eroded the combat readiness of the fleet, forcing the Venezuelan military to rely more heavily on the Sukhoi Su-30. *Venezuelan Ministry of Defense*

A Pakistani F-16 at Nellis AFB, Nevada for Exercise Red Flag, 2016. An unlikely Cold War ally, Pakistan received its first F-16s in the 1980s, whereupon they scored a handful of impressive aerial victories against intruding Soviet aircraft during the Soviet-Afghan War. *US Air Force*

Aggressor F-16s in flight over the Joint Pacific Alaskan Range Complex during an iteration of Red Flag Alaska, July 2009. Red Flag Alaska is a Pacific Air Forces-directed training exercise that simulates real aerial combat for American and Allied air force pilots. These F-16s belong to the dedicated aggressor units who play the role of enemy bandits against participating aircraft. *US Air Force*

A Royal Thai F-16 in flight during joint maneuvers over Hawaii, 2011. Thai F-16s saw action during the 2025 Cambodian-Thai Border Crisis, attacking Cambodian ground forces. *US Air Force*

A Royal Netherlands Air Force F-16 over Afghanistan, May 2008. With the onset of the Global War on Terror, the mission in Afghanistan drew a multi-national presence, as NATO air forces contributed to the fight against Al-Qaeda and the Taliban. *US Air Force*

An F-16C, piloted by LTC Bradford Everman, commander of the 177th Fighter Wing (New Jersey Air National Guard) takes off from Bagram Airfield, Afghanistan, 2012. *US Air Force.*

A pair of F-16s from the 363rd Air Expeditionary Wing on the "hot ramp" at Prince Sultan Air Base, Saudi Arabia prior to a night mission on March 21, 2003 during Operation Iraqi Freedom. *US Air Force*

An American F-16 in flight over Iraq, June 2008. *US Air Force*

Chapter 4:
Grey Zone Falcons

Following the end of Desert Storm, the skies over Iraq remained anything but peaceful. In the uneasy years that followed, two missions—Operation Northern Watch and Operation Southern Watch—defined a new chapter of American air power: The No-Fly Zone era. For more than a decade, these missions demanded constant vigilance from Allied air squadrons as they flew daily patrols over Iraqi airspace, deterring further aggression from Saddam Hussein.

This was not a war of sweeping offensives or "shock-and-awe" campaigns. It was a long-term reconnaissance mission, fraught with tedious air patrols punctuated by moments of sheer terror.

And at its core stood the F-16 Fighting Falcon.

The No-Fly Zones were born from the ashes of Desert Storm. After the ceasefire in February 1991, Saddam Hussein quickly turned his attention inward. In the north, he unleashed artillery and helicopter airstrikes against the Iraqi Kurds who had risen against him. In the south, a Shiite rebellion had been crushed with the same brand of brutality. In response, the US, UK, and France imposed aerial exclusion zones over the north and south of Iraq—initially to protect Kurd and Shiite civilians, but increasingly as a strategic tool to contain Saddam's military power.

Operation Southern Watch (OSW) began in August 1992 and covered airspace south of the 32nd Parallel (later extended to the 33rd). Its mission: Deny Iraqi aircraft the ability to operate freely in the south and to monitor compliance with UN resolutions. Operation Northern Watch (ONW) began in January 1997, replacing the earlier Operation Provide Comfort. Flying from Incirlik Air Base in Turkey, ONW patrolled the airspace north of the 36th Parallel. Here, the mission was similar: Protect Kurdish civilians and deter any Iraqi military buildup.

The No-Fly Zones weren't simply deterrent measures; they were active combat operations. And for the F-16 pilots, their fighter patrols could turn deadly in the blink of an eye.

Fast, agile, and multi-faceted, the Vipers were ideal for No-Fly Zone missions. Squadrons rotated in and out of theater every 90-120 days. For the aircrews, it meant long hours, unpredictable dangers, and an endless game of cat-and-mouse with Iraqi SAM batteries. Southern Watch was headquartered out of Prince Sultan Air Base in Saudi Arabia (and later, Al Udeid in Qatar), drawing Viper units from across the US—including Hill, Cannon, Shaw, and Luke Air Force Bases. In the skies of ONW, Incirlik Air Base became a second home for many European-based F-16 squadrons, particularly those from Aviano Air Base in Italy and Spangdahlem in Germany.

The daily mission was deceptively simple: Fly into the designated airspace, establish a presence, and observe. In reality, it was a high-stakes balancing act. Iraqi air defenses were still active, though not as strong as they had been prior to the Gulf War. Still, radars swept the skies, SAM batteries tracked incoming aircraft, and Iraqi MiGs occasionally pushed the boundaries. At any moment, the sky could erupt into combat. Pilots learned to fly with one eye on their mission systems and the other eye glued to their threat receiver. A sudden tone or flashing threat symbol could mean a radar lock; or worse, an incoming SAM.

It was during these No-Fly Zone missions when, on December 27, 1992, an American F-16 downed an Iraqi MiG-25. The thunder of Desert Storm had faded, but the skies over Iraq still crackled with tension. Saddam's fighters kept probing the latitudinal boundaries of the No-Fly Zones. And the 33rd Fighter Squadron—recently deployed from Shaw AFB to Saudi Arabia—stood ready to answer the call.

This was the unit's first rotation to OSW, but it hadn't been their first rodeo with Saddam Hussein. Two years earlier, the 33rd had deployed to Saudi Arabia during the buildup to Desert Storm, but most of those veterans had since rotated out. Indeed, by the fall of 1992, only three combat veteran pilots remained in the unit. The rest were green and hungry for action.

Their intel briefs warned them early: The Iraqi Air Force had been more active than usual. Local AWACS had observed Iraqi fighters breaching the No-Fly Zone before hastily retreating across the 32nd Parallel. Nearly all of these border incursions happened in the early morning hours, or when no US fighters were airborne. As of yet, there had been no provocations involving American F-16s, but the message was clear: The Iraqis were trying to elicit some kind of response. The airspace may have been quiet, but it wasn't safe.

On the morning of December 27, Lieutenant Colonel Gary North was leading a flight of F-16s on a typical Southern Watch mission—patrolling the No-Fly Zone for approximately 30-45 minutes before returning to the airbase at Dhahran. North was one of the few pilots in the squadron who had fought in Desert Storm. He knew the terrain. He knew the enemy. And he wasn't impressed. "Sometimes a flight would be ordered to overfly a specific area and observe any unusual ground activity." The main purpose of these flights was to let Saddam Hussein know that coalition aircraft were still on guard.

During an aerial refuel, North and his comrades monitored a transmission between the AWACS and a nearby flight of F-15s. "An Iraqi MiG-25 had crossed the border into the No-Fly Zone," he recalled, "flown within lethal range of the F-15s, and was speeding north to safety with the F-15s in hot pursuit." One of the F-15s had visually identified the bogey as a MiG-25 and requested permission to fire. By the time he received that clearance, however, the MiG had already retreated back across the 32nd Parallel.

After North and his wingman topped off on fuel, they sped into the No-Fly Zone, occupying their designated flight pattern while staying alert for any lingering MiGs. Minutes later, AWACS alerted North to a high-speed contact that had just breached the No-Fly Zone, 30 miles west of their position.

The F-16s vectored to intercept; but this MiG, too, retreated northward.

Shortly thereafter, AWACS reported yet another incoming bandit. This time, when the F-16s vectored to intercept, North's radar receiver indicated he was being tracked by a SAM battery. Yet, just as before, this MiG disappeared across the 32nd Parallel, and the SAM indicator fell silent.

By now, it was clear what the Iraqis were doing. They weren't just testing the airspace. They were sending their MiG-25s to lure the Americans into a pursuit, hoping to bring them within range of the SAM radars below. It was a proverbial "SAMbush," as the Americans would call it.

North knew the game now. And he wasn't going to play it on their terms.

As he resumed his patrol pattern, AWACS alerted him once again. This time, a radar contact had entered the No-Fly Zone on an eastbound vector, headed straight towards him. Determined not to let this MiG escape to its latitudinal sanctuary, North and his wingman maneuvered to "bracket" the incoming bandit. This maneuver essentially trapped the MiG within their immediate airspace; it couldn't retreat to the 32nd Parallel without a fight. As North recalled: "Someone was going to die within the next two minutes and it wasn't going to be me or my wingman."

North requested permission to fire, having identified the bandit as a MiG-25, while directing his wingman to jam the Foxbat's radar and communications. When North heard the transmission from ground control, "BANDIT-BANDIT-BANDIT. CLEARED TO KILL," he fired his AIM-120 missile, pounding the MiG at a distance of 20 miles within the No-Fly Zone. "The nose and left wing broke apart instantly, and the tail section continued into the main body of the jet for one final huge fireball."

It was the first confirmed air-to-air kill for an American F-16 and the first kill for an AIM-120 missile in combat.

Meanwhile, in the skies over the Northern No-Fly Zone, coalition pilots patrolled the fragile frontier between humanitarian protection and hostile provocation. Still under the auspices of Operation Provide Comfort (the predecessor to ONW), British and American pilots had been enforcing the Northern No-Fly Zone since April

1991, deterring Iraq from any further aggression against Kurdish civilians.

In December 1992, almost simultaneously with the 33rd Fighter Squadron's OSW deployment, the 23rd Fighter Squadron deployed from Spangdahlem, Germany to Incirlik Air Base for their inaugural mission into the Northern No-Fly Zone. Throughout Desert Storm, the 23rd had flown F-16s and F-4Gs in the "Wild Weasel" operations. After the war, the squadron was converted into a pure F-16 unit with a primary mission of Close Air Support. However, the SEAD requirements of the No-Fly Zone mission once again brought the F-4Gs and F-16s together as they paired up into hunter-killer teams.

Just like their brethren along the OSW front, F-16 pilots in the Northern No-Fly Zone found themselves in a deadly game of cat-and-mouse with the Iraqi Air Force. By late 1992, the tactics and tempo of Iraqi aircraft operating along the 36th Parallel had become more aggressive, facilitating the same "SAMbush" tactics that had cropped up in the OSW sector.

The coalition's response, however, was equally aggressive. Whenever a SAM radar was detected within the No-Fly Zone, Wild Weasels were quick to engage it. Just as they had done during Desert Storm, the F-16s and F-4Gs carried the AGM-88 HARM missile for SEAD operations, with the F-16s often carrying CBU cluster bombs for follow-on kills.

On January 17, 1993 (the second anniversary of Desert Storm), Lieutenant Craig Stevenson was a sharp, young Viper pilot assigned to the 23rd Fighter Squadron. Thus far, he had accrued more than 450 flight hours in the F-16C, and was determined to become a flight leader. For that, he needed a "checkride"—a dissimilar two-ship sortie with something other than an F-16. Technically, these certifying checkrides weren't supposed to happen in combat.

But combat had a way of rewriting the rules.

Thus, on the morning of January 17th, Stevenson climbed into his Viper for what was supposed to be a routine qualification flight. His wingman? A battle-hardened crew in an F-4G Phantom: Lieutenant Colonel Steve Heil and Captain Rich Piercey. Together, their job was to cover a pair of RAF Jaguars flying recon over suspected SA-6 sites in northern Iraq. Stevenson would lead the mission; the F-4G would handle the radar triangulation.

The first leg went by the book: Air refueling, ingress, and a sweep of the recon zone. The F-15s assigned to provide air cover had returned to base, so Stevenson's flight picked up the slack, now pulling double duty as both SEAD and CAP. And they were armed for it: Sparrows on the F-4, AMRAAMs and Sidewinders on Stevenson's F-16.

Then the tone changed.

Stevenson's radar lit up. It was a single contact, lifting off from a hostile airbase. Then he remembered the briefing: Iraqi aircraft from this base had a history of playing games along the No-Fly Zone.

But this wasn't a bluff.

Stevenson played it smart: He turned north, letting AWACS get a full read on the bogey. Then, he circled back, getting radar lock on the now-confirmed bandit. It was a MiG, climbing fast, heading north, right towards the No-Fly Zone.

AWACS then issued the call: "Commit."

Stevenson dumped his tanks and slammed into afterburner. The Viper surged, engine roaring like a beast unchained. Diving low, he dropped to 4,000 feet, trying to blind the MiG's radar by forcing it to look down, where its systems were the weakest. But the Iraqi pilot kept climbing, unaware or undeterred, reaching 17,000 feet. Both jets were now moving at Mach 1.2.

As the MiG crossed the 36th parallel—violating the No-Fly Zone—AWACS issued its final clearance: "Weapons Free."

The Iraqi pilot suddenly yanked hard right, reversing course and diving fast in a classic evasion maneuver. The quick maneuver broke Stevenson's radar lock. But that didn't matter: He reacquired the MiG visually at 6,000 feet, now in a tail chase. At five miles out, Stevenson was dangerously close to the edge of the AMRAAM's effective envelope. But he had speed. The MiG was running; and Stevenson was hunting.

Then fate stepped in.

The MiG pilot, seeing no escape, pulled hard left, turning back into the fight. It was a fatal mistake.

Stevenson pressed the trigger. But then, nothing. His left-wing AMRAAM stood cold and unresponsive. In the haze of speed and adrenaline, he switched to the AMRAAM on his other wing. Luckily, this right-wing missile roared off the rail, streaking into the sky like vengeance made manifest.

As the missile went active, Stevenson banked hard north, peeling away from the encroaching SAM threat below. He looked back.

A fireball. No parachute. No lingering doubts.

And he didn't linger as well. Fuel was running low. He made straight for the aero-tanker, sweating bullets as he lined up for refueling with one live AMRAAM still hanging on his wing. The Phantom crew, meanwhile, had remained behind, keeping an eye on the wreckage. Even as

Stevenson flew back, the smoldering smudge of the MiG was still curling skyward from the ground below. At first, there were whispers: Some thought it was a MiG-29. The truth? It was a MiG-23, verified later. But it didn't matter. A kill was a kill.

At various times throughout the engagement, Stevenson and his F-4G wingman overheard AWACS barking commands to another flight over the Guard radio frequency.

"Turn north immediately! You're in Syrian airspace!" the AWACS bellowed to the other flight.

Stevenson and the F-4G crew wondered whom the AWACS had been speaking to. They knew the AWACS couldn't have been speaking to them—they were using different callsigns and were nowhere near the Syrian border.

The culprits? A flight of F-15 Eagles, racing to the scene after hearing Stevenson's fight on the radio. "They thought somebody might need help," he recalled. They had taken a shortcut 100 miles through Syria, trying to reach the party in time.

But they were too late.

The kill was done. And it belonged to a young lieutenant on his flight lead checkride. In a brilliant stroke of fate, his flight lead upgrade mission had turned into the F-16's second air-to-air victory in US service.

Craig Stevenson had passed his flight lead evaluation…in the most definitive way possible.

For years after the Gulf War, American F-16s carved the same contrails across the skies of Iraq, policing the invisible borders of the No-Fly Zones. Day after day, these missions blurred together—the same routes, same checkpoints, and the same empty radar screens. By the late 1990s, the monotony was having a corrosive effect on the pilots' morale. Many of them began to joke that they were living the script of *Groundhog Day*, the 1993 film wherein the main character stays trapped in a single day that loops endlessly. With the Iraqi Air Force falling apart and offering no real fight, many pilots felt they weren't guarding the skies so much as burning holes in them, turning high-performance F-16s into very expensive fuel-burning noisemakers.

The pattern of these No-Fly Zone patrols continued uninterrupted until December 16, 1998, when President Bill Clinton authorized Operation Desert Fox—a four-day bombardment of critical Iraqi targets. By the fall of 1998, tensions had risen once again between Saddam Hussein and the international community. Iraqi officials continued to obstruct UN weapons inspectors, creating a deep concern over the regime's alleged Weapons of Mass Destruction (WMD) program. The stated goal of Desert Fox, therefore, was to eliminate any sites capable of manufacturing or delivering WMDs. And it would be the largest airstrike against Iraq since the Gulf War.

From forward-deployed US bases in Kuwait, the coalition marshaled its air power quickly. Among these combat-ready aircraft were several F-16s from squadrons representing the 20th and 388th Fighter Wings. The Vipers would join a broader air coalition of more than 300 combat and support aircraft, collectively flying more than 600 sorties throughout the 70-hour campaign.

In the days leading up to Desert Fox, the F-16 squadrons underwent extensive planning and mission preparation. Intel specialists collected data on Iraqi air defenses, potential locations for WMD production, and suspected SAM sites. This information was refined in tactical briefings, where pilots studied digital maps and the anticipated threat scenarios. Coordination with other coalition aircraft and naval units was essential—for the F-16 would be operating in concert with RAF fighters, US naval air squadrons, AWACS, and heavy bombers.

Mission planning revolved around precision and speed. According to one pilot: "Each sortie had its own fingerprint. No two nights were the same. We were handed a new threat matrix, new weather conditions, and sometimes new rules of engagement right before refueling." The F-16s designated for strike missions underwent final weapons checks by ground crews as pilots reviewed contingency procedures for jamming, kinetic evasion, and rerouting. The support teams—including ground fuelers, munitions techs, and avionics specialists—worked under floodlights, sometimes in sand-choked air, ensuring every system would endure the rigorous operational demands of working in a desert environment.

Although the F-16 was not the headliner aircraft of Desert Fox, it nevertheless played a critical role in supporting the B-1 and B-52 airstrikes. American F-16s also flew night missions alongside the hardy A-10 Warthogs, launching from Ahmad al-Jaber Air Base in Kuwait. These sorties helped suppress enemy defenses, facilitate target designations, and establish air dominance for follow-on strikes.

The F-16 mission profiles tended to follow a similar pattern: A two-or-four-ship formation would power up from their runways in Kuwait, intercept hostile radars, disable SA-6 batteries, and escort bombers into their designated target zones. Most of the Desert Fox missions were flown at night, thereby facilitating the F-16's nocturnal fighting edge. By 1998, the *Block 40/42* F-16s had been equipped to operate at night and in low-visibility using the LANTIRN navigation and targeting pod. LANTIRN allowed single-seat fighters to perform low-altitude, terrain-following ingress maneuvers and to self-designate targets for laser-guided bombs. In other words: LANTIRN had turned the F-16s into effective night attack platforms, enabling low-level penetration, target acquisition, and laser guidance without a second crewmember.

SEAD missions followed much of the same playbook as they had during Desert Storm. AGM-88 HARMs were still the weapon of choice, and the mission now belonged exclusively to the F-16CJ (the SEAD-specialized variant derived from Block 50/52). These SAM-hunting Vipers could let a single pilot locate and prosecute hostile radar emissions. As such, the newer F-16CJ had finally replaced the older two-seat F-4G Wild Weasel, thus providing the Air Force with a single-seat, highly-agile hunter for Iraqi radar sites.

On other occasions, during the mixed package flights, F-16s would provide overwatch for bomber ingress. They flew in low, streaming images and targeting data back to their command centers in real time. Viper pilots thus became the conduits of precision, illuminating target coordinates, providing cyclic intel feedback, and facilitating tighter strike coordination.

The opening sorties of December 16-17 were marked by fast, coordinated strikes against priority targets across Iraq. Pilots recalled the unmatched tension of the first wave: They skimmed low and fast, hugging the earth to hide beneath the enemy's radar horizon. In the cockpits, their world had been reduced to glowing green symbologies, pulsing threat receivers, and the steady hum of oxygen through their facemasks. Iraqi air defenses—repositioned and reinforced—waited ahead like unseen predators. Mission planners had warned that the Iraqis' mobile SAM batteries had been shifted to create new kill zones. Every second of ingress required mental recalculations, adjusting approach vectors on the fly and probing for the smallest seams in the enemy's defensive weave.

Egressing, however, was a true race for survival. As soon as the bombs left their rails, the F-16s rolled hard, engines screaming at maximum military power as they cleared the target area. Iraqi air defenses responded in kind, with streams of tracer rounds and the stabbing arc of angry SAMs. Defensive maneuvers, irregular climbs, and chaff releases became routine, aided by the counter-air tactics borrowed and refined from previous operations in the Gulf.

By December 19, Operation Desert Fox had been declared a success. Nearly 85% of the identified targets had been destroyed, all without the loss of a single Allied aircraft. The successive wave of F-16 strikes had neutralized air defense radars, communication nodes, and ammunition dumps, leaving visible craters and destroyed infrastructure in their wake—all confirmed by post-strike reconnaissance photos. Though the F-16s may not have carried the bulk of Allied firepower, their presence gave the air campaign its operational edge. The Viper's agility, multi-role flexibility, and SEAD ferocity ensured that heavy strike assets (like the B-1s and B-52s) could operate with reduced risk. In total, coalition aircraft struck nearly 100 targets across seven categories: Air defense systems;

Command & Control centers; WMD-related facilities; Republican Guard units; airfields; and economic infrastructure.

While broader narratives often highlight bombers and cruise missiles, the proliferation of multi-role fighters like the F-16—equipped with HARMs and night-fighting capabilities—reaffirmed the importance of adaptable air power. The relatively short-notice deployment of these F-16s to Southwest Asia showcased the Air Force's agility in force projection. Following Desert Fox, the US would resume its regular ONW/OSW missions until the start of Operation Iraqi Freedom in 2003.

Although Desert Fox lasted only four days, it did not mark the end of hostilities within the No-Fly Zones. In the final week of December 1998, Baghdad declared the No-Fly Zones void and ordered air defenses to engage coalition aircraft. The result was a sudden spike in SAM launches and a rapid shift in coalition posture from passive patrol to active suppression. The change was quick and brutal: When Iraq fired, Allied aircraft answered with HARMs, precision bombs, and a willingness to take out entire elements of the air defense web, not just the launcher that had fired the missile.

Prior to Desert Fox, the lack of enemy activity (along with the *Groundhog Day* analogies) had slowly given rise to a culture of risk aversion within the ONW/OSW framework. Pilots had been encouraged to avoid and escape hostile threats rather than return fire. Brigadier General David Deptula, who commanded the Combined Task Force-Northern Watch out of Incirlik during Desert Fox, later said the new posture removed the old culture of risk aversion.

In other words: Once fired upon, a pilot could respond

immediately.

That change, though small on paper, became pivotal in practice.

What followed from December 1998 to March 1999 was a month-by-month attrition campaign against radar stations, control vans, SAM batteries, and anti-aircraft positions. The strikes were surgical and relentless: HARMs hunting radar sites, followed by GBU-12s and other precision-guided bombs to finish the job. The results were not always spectacular in the sense of making headline news. But when compiled over weeks, the data was irrefutable: Coalition airstrikes had decimated Iraq's air defenses and re-asserted control over the No-Fly Zones. Historians later counted the toll: hundreds of weapons expended and dozens of SAM radars destroyed or degraded in the months following Desert Fox.

By the time the ONW/OSW patrols gave way to the thunder of Operation Iraqi Freedom, the F-16 had flown tens of thousands of sorties over Iraq. From 1992-2003, the Viper had accomplished a number of milestones in its young service life with the US Air Force. It had fired the first AMRAAM in combat, suppressed countless SAM sites, and served as both shield and sword for coalition forces. More than any other fighter, it became the symbol of the No-Fly Zones—a tireless sentinel that never let Saddam forget he was being watched.

In the end, the Viper's postwar mission over Iraq was more about endurance than glory. Unlike the aerial blitzkrieg of Desert Storm or the decisive strikes over Kosovo, the No-Fly Zones were a marathon—twelve long years of vigilance. Pilots rotated in and out, with generations of air and ground crews cutting their teeth on real-world missions that were routine, yet fraught with danger. For them, the desert sky was a crucible where patience met sudden violence, and the line between war and peacekeeping could blur within a matter of seconds.

When the last No-Fly Zone patrol ended in March 2003, the F-16 had already written its legacy over Iraq. It was the workhorse of containment, fighting a war measured not in territory taken, but in the freedom denied to an adversary below. In that long and winding mission, the Viper proved itself not just as a multi-role fighter, but as an unyielding guardian of the skies.

The Balkans

Meanwhile, in the aftermath of Desert Storm, the Socialist Federal Republic of Yugoslavia began to fracture along ethnic and religious lines. Yugoslavia, a "non-aligned" Communist state bordering the Adriatic Sea, had been the dominion of President Josip Broz Tito. Following Tito's death and the collapse of Communism in Eastern Europe, however, the ethnic and religious groups within Yugoslavia (Bosnians, Serbs, Croats, et al) began jockeying for independence. These revolutionary movements, however, soon devolved into a civil war, whereupon NATO and the UN eventually intervened.

Like their counterparts in the former Soviet Union, the Yugoslav Air Force saw much of its inventory absorbed into the successor states. Unlike the Soviet Union, however, these former Yugoslavian entities turned their weapons against each other.

When the Yugoslav Wars began, Serbian forces effectively took control of the air assets in Belgrade. With the Serb-dominated federal government seeing itself as the true heir to President Tito's legacy, they began launching airstrikes against Croatian targets in August 1991.

These domestic air attacks quickly ended, however, with the onset of NATO's intervention. Beginning with Operation Sky Monitor in November 1992, followed by Operation Deny Flight in March 1993, NATO aircraft gradually increased their role in the skies over Yugoslavia. By the summer of 1993, US and NATO aircraft could fly fully-armed, Close Air Support missions in support of UN peacekeepers.

But to the pilots themselves, the "peacekeeping" mission felt more like a sightseeing tour. Indeed, the heart-pounding combat sorties of Desert Storm were few and far between. US Air Force pilot Lieutenant Colonel Michael Arnold, summarized the collective frustrations shared by most NATO flyers at the time: "Most of the missions were very benign and very dull," he said. "We enforced the No-Fly Zone, so we would go up and try to engage anyone who was not supposed to be flying, but of course, we never saw anything."

Operation Sky Monitor began in 1992 with all the political sharpness of a putty knife. The UN had declared Bosnian skies off limits to military flights, but NATO's orders were clear: Observe, Record, and Report. Do not fire unless fired upon.

For the American F-16s at Aviano Air Base in Italy, it was a mission of restraint. AWACS would send contact reports, and the Vipers would vector to confirm what they saw. The mission logs populated with call signs and coordinates, while gun cameras caught the fleeting silhouettes of transport planes or light attack jets.

But the rules were clear: No engagement.

One Aviano pilot called it: "A mission no one wanted to take." Long hours at higher altitudes; cold intercepts; and the knowledge that whatever you found, you couldn't touch. The cockpit became a place where boredom and frustration shared co-equal residency.

By the winter of 1992, however, the UN's patience with mere observation was growing thin. The Serbian Air Force was still flying. UN peacekeepers on the ground saw aircraft strafing villages and bombing towns in broad daylight. In March 1993, the UN issued Resolution 816, authorizing "all necessary measures" to enforce the No-Fly Zone. Thus, on April 12, 1993, Sky Monitor gave way to Operation Deny Flight.

The watchers had become the hunters.

The shift was almost immediate. F-16s were now flying enforcement CAPs. Loadouts changed: AIM-9s and AIM-120s now hung from the wings, with a complementary set of external tanks. Still, the leash remained short: NATO-UN protocols meant political clearance had to travel halfway around the world before a missile could leave its rails.

It was a maddening paradox. The mission was to keep enemy aircraft out of Bosnian airspace. The pilots had the skill, the jets, and now the legal authority. But clearance could take minutes... or never come at all. In the cockpit, this meant shadowing a hostile aircraft at gun range, finger on the trigger, and knowing that the only thing preventing your shot was a bureaucratic decision from Brussels or New York.

One veteran from those early Deny Flight patrols described the tempo as: "Long stretches of monotony broken by thirty seconds of chaos." Every mission seemed to follow the same script: A radar flash; a vector from the AWACS; and a pilot who was seconds away from getting a good tone before the call came back: "Abort; no clearance." The target turned away, and the adrenaline devolved into frustration.

Throughout 1993, F-16s orbited Bosnia day and night. They escorted AWACS, shielded reconnaissance flights, and intercepted the occasional rule-breaker. They learned the geography of the war by heart: The Drina Valley, the Sarajevo bowl, and the ridgeline crevices where mountains hid SAM batteries. Pilots flew with an awareness that the Serbian SA-6 and SA-7 launchers were not relics. "We knew the radar was painting us," one recalled years later. "We just hoped the ROE [Rules of Engagement] would let us do something before they pulled the trigger."

In the cold light of the Adriatic mornings, the Vipers often returned to Aviano with bug-smeared canopies and pilot fatigue—not from combat, but from the endless cycles of "load, fly, and stand guard."

They were the visible edge of NATO's military might, but were still bound by its caution.

By the end of 1993, Deny Flight had stopped some flights, deterred others, and given NATO its first sustained combat presence over Bosnia. It also highlighted a bitter truth: Enforcing the Bosnian No-Fly Zone was as much about politics as it was about the pilots. Over the next two years, the F-16s would find themselves pushed closer to the war—and eventually, into it. But in 1993, they learned the art of waiting with weapons at the ready.

By February 1994, Operation Deny Flight had been patrolling the No-Fly Zone for almost a year. But the mind-numbing, tedious nature of "peacekeeping," and the frustration of shadowing hostile aircraft without permission to engage was taking its toll on the pilots. By now, they could identify Serbian aircraft on sight. They could distinguish the colors of their wing flashes and the shapes of their faded roundels. But they also knew that these Serbian jets couldn't be touched.

However, that abruptly changed on the morning of February 28, 1994.

Six Serbian J-21 Jastrebs lifted off from Udbina Air Base, running low across the ridgelines towards central Bosnia. Their mission was simple: Destroy the Bosnian armaments factory at Banja Luka. It was a deliberate, public violation of the No-Fly Zone. And this time, NATO was watching. British AWACS caught the bandits on scope as they crossed into the prohibited airspace.

Meanwhile, a two-ship flight of F-16s from the 526th Fighter Squadron was on patrol south of Mostar. Led by Captain Bob Wright and his wingman, Captain Scott O'Grady[3], the pair had grown accustomed to the monotonous and uninspiring routine of Bosnian air patrols. Their mission was marked by boredom and a sense of obligatory impotence.

But those feelings were about to change.

The AWACS message was cold and clinical: Vector to intercept. If the Serbian pilots ignored orders to land or vacate the area, they would be fired upon.

The two F-16s rolled hard onto their vector, making a

---

[3] O'Grady would make headlines the following year when he was shot down by a Serbian SAM. He ejected over Bosnia and evaded the enemy for six days before being rescue by a Marine Corps heliborne recovery team.

beeline dash towards the J-21s as they climbed out of the low ground. The Jastrebs, old Yugoslav attack planes, were small and fast at lower altitudes; they had no modern warning systems and flew low to hide behind terrain. But on that day, the F-16s had the technology and the Rules of Engagement to ensure victory.

Wright's first shot was an AIM-120 fired in radar mode at the trailing Jastreb. The missile found its mark and the plane spiraled downward in flames near the village of Bratstvo. He followed with AIM-9 Sidewinders against two more Jastrebs as they tried to thread the valleys and escape the fight. Both shattered under the quick, heat-seeking detonations of the vengeful AIM-9. O'Grady, meanwhile, engaged one of the fleeting Jastrebs, but his missile petered out before it could hit the target.

Because both men were now running low on fuel, Wright and O'Grady had to leave, but they were replaced on station by a second pair of F-16s from the 526th. That second pair, which included Captain Stephen Allen, caught up with the remaining bandits, with Allen scoring his own AIM-9 hit, downing another Jastreb at 6:50 AM.

But at first, it appeared as though the Serbian bandit was going to escape.

As Allen recalled: "I selected the AIM-9 Sidewinder missile from my left wingtip station. I mashed down on the pickle button, and the missile shot right off the rail. The missile plume was easy to track in the dim light." Visually tracking its trajectory, the AIM-9 seemed to be on a perfect glide path. But then, halfway towards the target, Allen suddenly lost sight of the missile plume.

Had the missile vectored off?

Had the rocket motor burned out?

He didn't know.

"I mentally counted down the few seconds it should take to get there, but nothing happened." He thought he might be experiencing "time dilation," a sensation wherein time seems to slow during moments of terror or excitement, and he waited a few more seconds. Still, no fireball. Undeterred, he armed the second AIM-9 and fired it off just as he'd done with the first. "This time, however, I was rewarded with a tremendous fireball 4,000 feet off my nose," he said.

But the sheer size of the fireball is what shocked him the most.

For being such a small aircraft, the J-21 yielded an enormous explosion when hit by the AIM-9. "The fireball was still expanding as the debris descended into the treetops and began to tumble across the snow-covered ground," said Allen. By the time the smoke cleared, four Serb light-attack jets lay in pieces. It was the first air-to-air engagement in NATO's history. Years later, Allen reflected on the engagement, saying: "I bear no animosity towards the Serbian pilots who flew that day. In fact, I sort of admire them, all politics aside. They executed a well-planned, low-level ingress in less-than-optimal conditions, and conducted a successful attack."

Wright, O'Grady, and Allen later described the mission as a concentrated burst of work that felt like years of training compressed into three minutes: Target acquisition, missile launches, threat avoidance, and then a sudden return to routine. Wright's later reflections, as recorded by Aviano base historians and cited in official Air Force histories, read as the afterimage of a crackerjack fighter pilot doing his job under a political leash. It was a mission no one wanted and a mission that seemed to lack purpose. But at the moment of truth, he did what all fighter pilots train to do: He engaged the enemy.

The political geometry, however, is what truly shaped these early years of the Bosnian mission. Deny Flight existed in a space carved by UN resolutions and NATO politics. The authority to fire was never automatic, and the dual-key arrangements of the time meant that legal clearance could be slow. In this case, AWACS and theater command granted authority to engage when the Jastrebs dropped ordnance inside of Bosnia—a clear breach of the No-Fly Zone. The response was sharp: An immediate, visible lesson about enforcement. Within weeks the Serbian Air Force curtailed its fixed-wing sorties, proving that NATO's threats had teeth.

Official US tallies credited Wright with three kills and Allen with one kill. But some accounts have differed regarding the number of aircraft lost that day. For example, Serbia's Ministry of Defense and other regional accounts later listed *five* Jastrebs lost on February 28. These computational differences, however, do not change the essence of the Banja Luka Incident: NATO had fired, and Yugoslavian jets had fallen from the sky.

After the mission, the world reacted in small but sharp ways. For the airmen at Aviano, the engagement had erased the monotony of Operation Deny Flight. Rules had been enforced; pilots had been called upon; and NATO had chosen decisive action over passive observation. For the pilots themselves, it was more complicated: The elation at having done their job, mixed with the heavier feelings that arise whenever a training exercise becomes a funeral pyre for enemy aircraft. Steve Allen, upon landing at Aviano, was marshaled by his commander onto the nearest telephone for an interview with the press. Although Allen was in no mood to entertain prying journalists, he reluctantly obeyed the order, and took the receiver for his tele-interview. One reporter asked him if he felt like he'd been "shooting fish in a barrel." Allen found the remark insensitive, implying that the F-16s had been bullying a weaker air force. "I told him to interview the victims of the bombing raid," Allen recalled, "and ask *them* how it felt to be fish in a barrel." For NATO, the Banja Luka Incident was a lesson in how quickly the skies over a contained mission could become a proving ground of political willpower.

Still, the war didn't shrink from the sky. Throughout late 1994 and into 1995, Serbian forces adapted. They shifted from fixed-wing strikes to artillery sieges and harassment fire. The No-Fly Zone still mattered, but Deny Flight began to evolve into something more dangerous: Close Air Support and deliberate strikes. Now, F-16s were tasked to escort attack jets into hostile territory, and sometimes dropping their own ordnance on artillery positions threatening UN troops.

As the mission grew, so too did the risks.

The Bosnian Serb air defenses weren't theoretical—they were armed to the teeth with SA-6 SAM batteries. Pilots knew the missile envelopes; they knew where the radars hid in the valleys; and they knew that one lapse in situational awareness could mean a white-hot streak in the sky followed by a forced bailout over hostile terrain. In the summer of 1995, such an incident would befall an American F-16 pilot.

The sky over Bosnia was a cold gray canvas on the morning of June 2, 1995—electrified with the hum of turbine engines and the tension of a No-Fly Zone that was about to be tested. Captain Scott O'Grady, mission call sign Basher 52, keyed the radio and sent a quiet, routine check to his wingman, Captain Bob Wright, high above the mountains.

The pair were still riding high from their involvement in the Banja Luka Incident a year prior. That brief dogfight had shattered the cruel monotony of their daily peacekeeping patrols. During that engagement, Wright had downed three Serbian J-21s, while O'Grady had engaged a fourth aircraft that managed to slip away.

Today, however, there seemed to be no such excitement on the horizon.

In fact, today's patrol had been nothing but routine. NATO flight duty was *supposed* to be that way. But beneath the calm sky, something lethal was waiting.

Suddenly, the silence broke as O'Grady's headset erupted with the dreadful blare of a SAM warning. The instrument panel screamed: He was being painted by an SA-6 battery hiding in the valley below. The SAM crew had been waiting for his Viper to pass through their blind spot—and they fired their missiles with perfect timing.

Inside the cockpit, O'Grady tore the throttle into a defensive split as two angry SAMs erupted into his path. The first missile thundered right between his and Wright's aircraft, splitting their formation with a deafening roar as its contrail sliced into the air.

Having narrowly missed the first SAM, O'Grady bucked the F-16 into a tight, evasive turn...but it was too late.

The second missile scored a direct hit.
O'Grady felt the impact with a jarring jolt, followed by the white flash of a detonating fuse. His instrument panel went red as the ailing Viper disintegrated underneath him.

With the F-16 pitching downward into a violent

shudder, there was only one option left: Eject.

The Viper's canopy flung open with a metallic clang and, with another quick jolt, he was floating downward into hostile territory. His aircraft vanished somewhere over the horizon, cratering into the ground with a violent medley of smoke and billowing flames. Bob Wright saw the explosion, but never saw the parachute.

Meanwhile, Scott O'Grady touched down in the wilderness near Mrkonjić Grad, deep in Serbian-held territory. Surrounding him was a world of trouble: Serbian checkpoints, paramilitary patrols, and the savagery of a war that paid little mind to the Geneva Conventions. Yet, the terrain offered him a cruel sanctuary. There were rocky hills, thick forests, and plenty of places to hide from angry Serbs.

He took stock of his gear—a survival vest; sidearm; and a 29-pound survival bag with compact rations and a radio. O'Grady remembered from his survival training that "running usually led to capture." So, he made the hardest decision a downed aviator could make: *Stay put. Don't move during the day. Move only at night.*

For six days, he lived by that rule.

By the daylight, he laid still. By night, he moved. Still, survival was slow and animalistic. His food ran out quickly, forcing him to survive by foraging, eating ants, chewing leaves, and sopping rainwater into plastic bags from his survival kit. Years later, he explained how his survival training had been mundane until it mattered. Now, he was hiding, face-down while Serbian troops combed the woods, sometimes coming within mere feet of his position. He could see the shuffling of their boots and hear the cadence of their speech.

It was a true game of life-and-death.

For days, NATO aircraft scanned the area, listening for his distress beacon. For much of his ordeal, O'Grady's handheld radio was an unreliable thread to the outside world—sometimes it worked, sometimes it didn't. Then, in the small hours after midnight on June 8, he keyed a call that would change everything. A nearby AWACS relayed his voice; another American pilot answered. On the relayed frequency, he spoke the words that stopped the quiet panic within the Allied Headquarters:

"This is Basher 52. I'm alive."

That confirmed his identity and set in motion one of the most tightly coordinated rescue operations of the Bosnian Campaign.

From the amphibious assault ship USS *Kearsarge*, Marines of the 24th MEU loaded onto a pair of CH-53 Sea Stallion helicopters, escorted by AH-1W Cobra gunships and AV-8B Harriers. At dawn, the aerial rescue force rose in a fury of determination, slicing across the Adriatic Sea towards imminent danger.

Following the yellow smoke flare that O'Grady had fired to mark a clearing, the rescue team established a perimeter before the second insertion party even landed. Scott O'Grady emerged from the forest, sporting a limp-gaited yet triumphant stride toward salvation. The Marines ushered him aboard and, in a harrowing climb away from the Serbian air defenses, they barreled their way back across the Adriatic. Within minutes, O'Grady and his rescuers were safe aboard the *Kearsarge*.

Standing on the deck of that amphibious assault ship, Scott O'Grady was dehydrated, exhausted, and in the worst physical shape of his life. He had been wet, cold, and half-starved for days. Yet, his first words reflected the resolve of a man who had done what was necessary, and just wanted to go home.

Still, the shootdown and rescue had become a media sensation. Political leaders hailed it; the public saw it as a dramatic microcosm of the Bosnian intervention; and NATO found a harder edge to its mission. For the US servicemen at large, it was a nod to their sacred credo: "Leave no man behind."

O'Grady received a Bronze Star and the Purple Heart. But the quiet truth of those six days—the claustrophobic fear, the queasy decisions about what to eat, and the laser focus on staying invisible—remained with him. Standing before his comrades at Aviano Air Base, O'Grady reflected that the ordeal had been transformative. It was cold and agonizing, but also life-affirming. Over the course of six days, he had shed 25 pounds and experienced a drop in body temperature to near-hypothermic levels.

Yet afterwards, he called those six days "the most positive experience of my life." In his view, the experience had tested every ounce of his training, faith, and his will to live. His story is not only a tale of rescue, but a lesson on training, improvisation, and the small, stubborn mechanics of survival.

NATO's Southern European commander, Admiral Leighton Smith, summarized it best: "This is a tough hombre we're talking about. Whatever else he had, he had a lot of guts to go with it."

For several years thereafter, O'Grady's words and the images from that week remained in the public consciousness: The pilot who ate ants, hid like an animal, and then ran with a pistol into a hovering helicopter. In fact, his story became the inspiration for the 2001 film *Behind Enemy Lines*, starring Owen Wilson and Gene Hackman. But the most memorable lines from O'Grady himself are the simple ones he spoke at the moment of relief: "This is Basher 52. I'm alive;" and, once aboard the helicopter: "I'm ready to get the hell out of here." They are the blunt language of survival, and the reason the operation has been taught in survival courses and retold in histories of NATO's air campaign.

Two months later, the Allied peacekeeping mission in Bosnia took a serious turn. After a relentless string of Bosnian Serb mortar attacks on Sarajevo—most notably the Markale Market Massacre, which killed 37 civilians and wounded dozens more—NATO launched Operation Deliberate Force on August 30, 1995.

It would be the first large-scale, sustained aerial campaign in NATO history.

The airstrikes were designed to pressure the Bosnian Serbs into halting their attacks on UN "safe areas" and compel their return to the negotiating table.

On August 30, the dawn over Sarajevo came up like thunder. The 31st Fighter Wing, based in Aviano, had flown over the Balkans since the earliest days of Deny Flight, but now that posture had changed. At 2:00 AM local time, NATO officially commenced Operation Deliberate Force. Within hours, F-16s from the 510th and 555th Fighter Squadrons were airborne, munitions locked and sensors alert.

Flying through the battered clouds with missiles at the ready, Vipers cut through the heavy air of the Bosnian summer to find their targets. Their missions followed a textbook-level precision: Laser-guided bombs pummeled Serbian radars and ammunition dumps, supported by SEAD flights and AWACS guidance. The sonic thread of HARMs firing, the hiss of the wind passing over the pylons, and the muffled chorus of exploding bombs—it was the poetry of modern warfare.

During those few weeks, the F-16s from Aviano launched more than 1,600 sorties, delivering swift and unparalleled punishment to the Bosnian Serb enclaves. But these pilots also understood a deeper truth: The fury and tempo of their pinpoint airstrikes had motivated the Serbs to re-enter negotiations. Post-strike analyses concluded that these F-16s had played a pivotal role in "destroying numerous key military sites and forcing the withdrawal of heavy weapons from the exclusion zone around Sarajevo."

The aerial bombardment was deliberate, but not indiscriminate. Indeed, every LANTIRN pod, every laser-guided bomb, was tailored to exact coordinates. The flanks of the campaign moved with such precision that civilian losses were minimized (although not completely avoided). And every ground controller, sensor operator, and pilot understood a single truth: This was surgical air power in real time.

And with help from dedicated SEAD units like the 23rd Fighter Squadron, NATO would soon pummel the Serbian air defenses into total submission. The pilots of the 23rd had flown south from Germany a few days earlier, trading the pine groves of northwestern Europe for the jagged mountains of Italy's upper coast. The mission was simple on paper, but brutal in practice: Find the enemy's radars and kill them. Or, in the alternative, scare the Serbian batteries into silence. Such was the prelude to nearly every NATO airstrike of Operation Deliberate Force. It began in the darkness, with Wild Weasels taking the first step into Bosnian airspace.

Even before the mortar attack on Sarajevo's Markale market, air planners had spent months building target sets and mapping aerial routes. When Deliberate Force began on August 30, the first wave was a classical SEAD operation: Destroy anything on the battlefield that can "see" or "talk." That meant EW jammers in the stack, AWACS stitching radio calls, and Wild Weasels carrying HARMs into the fight. But the Combined Air Operations Center (COAC) soon began generating more sorties than NATO had anticipated, forcing a surge of additional SEAD assets into theater. In fact, the F-16CJs with HARMs Targeting System (HTS) were in such high demand throughout Deliberate Force, that the tempo of the air campaign quickened or slowed depending on how many Wild Weasels were available at any given moment.

The Bosnian Serb air defenses, meanwhile, clung to their patchwork of Soviet-era hardware: SAMs, air defense guns, and MANPADS, all stitched together by a radio network that knew the hills by name. The layout wasn't dense by Gulf War standards, but it was cunning. And the terrain was a defender's best friend: Folded ground to mask emissions, valleys to canalize incoming attackers, and ridgelines that turned radar horizons into jagged sawteeth. But just as the Wild Weasels had done in Desert Storm, their job wasn't simply destruction, it was also suppression. In other words: They had to make Serbian air defenses *hesitate*—forcing them to go dark at the wrong times, then parlay those seconds of doubt into minutes of sanctified airspace for Allied strike packages.

By the first morning, the pattern was set. A two-ship Wild Weasel element would cross the coast at medium altitude, sweeping the horizon with its HTS. If a radar painted them, the HARMs came off the rails, riding home on the radars' energy. If the enemy radars stayed dark, the Wild Weasels pressed in, bracketing suspected SAM sites with the threat of instant punishment, while the strike packages flew in behind them, engaging every priority target that held the Bosnian Serb front together.

The opening volleys battered Serbian air defenses in southeast Bosnia with a relentless fury. In fact, the tempo was so relentless that CAOC's operational staff could hardly keep up. New sortie requirements and minute-to-minute re-taskings became a daily occurrence. Indeed, a strike package briefed to attack one valley could be re-tasked on the flightline to attack a different valley, consequently shifting the flow of aerial tankers.

On September 1, however, politics forced a pause. The UN gave Bosnian Serb forces an opportunity to end the siege of Sarajevo and withdraw from the exclusion zone. But, four days later, the campaign resumed when Bosnian Serb forces failed to respond.

The second act of the air campaign hit faster. Although the weather was uncooperative, the SEAD pilots did their part—silencing enemy air defenses, holding corridors, and punishing ground targets. By September 7, NATO commander Admiral Leighton Smith reported that the Bosnia Serb air defenses had been significantly degraded. They were still a threat, but they had taken enough of a beating that the Allied strike force could now fly at a lower risk.

Over the next three weeks, the 23rd Fighter Squadron tallied a total of 224 sorties—missions that didn't make headlines or the evening news, but made everything else possible. Every flight was a gamble that the other side would blink first, and that the HARM's logic chain would be shorter than the SAM's.

One mission, etched into the squadron's lore, occurred on September 6, 1995. On that Wednesday morning, an

HTS-equipped F-16C, found its first dance partner of the day—a SAM launcher had just activated its radar. Considering the circumstances, however, it made no sense for the SAM crew to announce their presence when they did. In a fight where well-timed invisibility meant survival, the SAM operators had just painted a target on their own backs.

Perhaps they believed the cloud cover was enough to conceal them. Perhaps they had miscalculated the relative distance. Or perhaps they had more guts than good sense. But whatever the reasons, the SAM's radar broadcasted their position, prompting the F-16 to fire its AGM-88 HARMs...the first-ever HARMs engagement cued by the new HTS targeting system. As expected, the SAM launcher paid dearly for its mistake. But in that instant, the HTS had gone from promising technology to combat-proven. Back at the CAOC, the legitimacy of a now combat-validated HTS made planners more aggressive in threading packages into Bosnian airspace, and it widened the range of acceptable targets under the ROE. In sum, the HTS-guided HARM became its own example of a positive feedback loop: Success bred trust; trust accelerated tempo; and that higher tempo had compounded the effects of NATO's air campaign.

By mid-September, the target list was growing thin because the effects of the air campaign had been devastating. NATO aircraft had engaged 338 aim points inside 48 complexes with a precision ratio that would have astounded Vietnam-era tacticians. Roughly 70% of the weapons expended by NATO were precision-guided, and the kill rates showed it. Diplomats could now speak with a new kind of gravity. The air war wasn't total (and it wasn't intended to be) but it was coercive in the exact way its planners had envisioned: Destroy what matters most to the aggressors, limit their means of continuing aggression, and make every passing day of defiance more costly.

As Deliberate Force began to nudge the warring parties back into negotiations, there was little doubt that F-16 SEAD operations had helped shape the campaign's tempo and its strategic impact. Across every Allied squadron, the sortie generation rate during Deliberate Force quickly outpaced all pre-mission forecasts. Given the tactical priorities of the air campaign, the HTS-equipped Viper became a flow regulator for the entire operation. With a robust Wild Weasel network, NATO could launch larger, more frequent strike packages and accept more ambitious target sets. Without these F-16 Wild Weasels, air commanders would have assumed a greater risk of attrition, fratricide, or mission aborts under the ROE's demand for positive identification. And as the air campaign matured into its successive weeks, the demand surged for more F-16 SEAD support. As reported by a US Air Force study, the number of F-16 Wild Weasels in the air directly translated to how many strike packages could fly safely each day.

But SEAD's effectiveness wasn't measured only in terms of dead radars. It was also a function of "Dead Eye" gamesmanship. In other words, several SAM launchers had survived simply by staying dark. Every time a SAM battery withheld radar emissions to avoid HARMs, it forfeited battlefield awareness. That gap gave NATO strike aircraft their windows of opportunity. In this way, SEAD F-16s traded missile engagements for enemy uncertainty, and uncertainty buys time—time to ingress or egress unmolested by enemy fire; time to maneuver around bad weather; or time to coordinate with UN ground forces who often helped reshape target lists in real time.

Unwittingly or not, SEAD F-16s also mitigated points of friction within the Deliberate Force ROE. Given the complexities of the UN-NATO power structure, the theater air commander had to personally approve ground targets, and even specific aim points, as a means to prevent collateral damage. Understandably, the bureaucratic tedium would impose delays and frequent re-taskings. SEAD mitigated this friction by making dynamic re-routes survivable. A late change to a bridge farther north might re-draw the aerial tanker plan and force a strike package into unfamiliar valleys. But with the Wild Weasels in front, the strike package could still sally forward, even during inclement weather, and still engage their targets within parameters of the ROE. Without the SEAD-capable Vipers, those last-minute changes would have died on the ramp or risked catastrophic losses to Allied air squadrons.

Operation Deliberate Force ended on September 20, 1995. And the outcome was devastatingly clear: 3,515 missions flown; 708 precision-guided bombs delivered; and all without the loss of a single NATO pilot. A French Air Force Mirage 2000N-K2 had been shot down by a MANPADS missile on the first day of the campaign, but both crew members survived and were eventually released by Bosnian Serb forces.

In raw numbers, the 3,515 sorties from Operation Deliberate Force barely matched a single day's worth of combat during Desert Storm. But in the skies over Bosnia, every mission carried the weight of a headline and the scrutiny of a courtroom. Yet, the operation yielded significant results relative to its size. Designated strike package aircraft (including non-SEAD F-16s) flew nearly sixty percent of the missions in Deliberate Force, and delivered most of the precision-guided ordnance that was used during the campaign. But that efficiency depended on a persistent SEAD umbrella. By keeping air defenses suppressed, SEAD F-16s had essentially paved the way for NATO's air campaign. When the airstrikes paused from September 1-5, it was the SEAD-enabled striking power that made the restart so decisive in the eyes of both diplomats and belligerents.

Deliberate Force ended not with parades, but with signatures and handshakes. Having felt the wrath of NATO air power, the Serbs reluctantly returned to the negotiating table. Negotiators from Bosnia, Serbia, and Croatia gathered in Dayton, Ohio to craft an enduring framework for peace in the Balkans. The resulting Dayton Accords, signed on November 21, 1995, formally recognized the state of Bosnia-Herzegovina, consisting of two entities: The Republika Srpska, housing the Bosnian Serb population; and the Federation of Bosnia & Herzegovina, populated mostly by Croat-Bosniaks.

Deliberate Force had been 23 days of gunmetal skies; then NATO returned to business as usual. Operation Deny Flight resumed on September 21, 1995, terminating on December 20 after the Dayton Accords had taken effect. Still, the impact of Deliberate Force had been undeniable. Sarajevo, with the help of precision airstrikes, had entered into a new era of sustainable peace.

But while Bosnia and Croatia celebrated their postwar peace and independence, Serbia & Montenegro continued their political union, operating as the "Federal Republic of Yugoslavia." Consequently, the new Yugoslav Air Force was drawn from remnants of the legacy force that remained under Serbian control. At its peak, the neo-Yugoslavian air defense command had several dozen Soviet-era and domestically-built aircraft in service.

However, due to the UN Arms Embargo, the Serb/Yugoslav Air Force began to deteriorate from lack of spare parts. Meanwhile, as the rump-state government was trying to assert itself as the successor to Tito's Yugoslavia, they soon faced an insurgency in the province of Kosovo. By 1998, the separatist movement had escalated into a full-scale, high-intensity conflict. To make matters worse, Yugoslav President Slobodan Milosevic had initiated his own genocidal campaign against the Kosovar people, once again precipitating NATO's intervention in the Balkans.

NATO—already having a foothold in the region from the IFOR/SFOR peacekeeping mission in Bosnia—was now determined to drive the Serbians from Kosovo on the grounds of humanitarian intervention. Thus, on March 24, 1999, the air campaign for Kosovo began under the banner of Operation Allied Force—a 78-day air campaign to halt the ethnic cleansing in Kosovo and break Milosevic's forces without putting NATO boots on the ground.

It was a war of radar scopes and radio silence; of stealth jets and very public consequences; of fighter patrols meant to turn radar blips into confirmed kills. According to NATO parameters, Allied Force would have three objectives: (1) Ground interdiction to keep Serbian forces out of Kosovo; (2) Close Air Support as needed; and (3) Establishing air superiority over the region. The third objective, naturally, implied shooting down any hostile

Serbian/Yugoslav aircraft. And from the first night until the last, the F-16 was everywhere: Sweeping, escorting, and killing enemy MiGs.

The NATO coalition was a tapestry stitched together from more than twenty flags. But the heartbeat of several missions came from a familiar silhouette: The F-16 Fighting Falcon. American F-16s hunted from above, relentlessly looking for SAM radars; Dutch and Belgian F-16s flew as partners in the Dutch-Belgian Deployable Air Task Force (DATF); while Norwegian F-16s held the northern gate, flying CAPs out of Italy. Each nation carried its own rules, its own politics, and its own customs. But their missions converged over the same kill boxes, under the same aerial tankers, and against the same Serbian forces that made headlines for all the wrong reasons.

By the time NATO air squadrons arrived in 1999, the violence in Kosovo had spiraled out of control. Diplomatic overtures seemed promising at first, but failed quickly. Milosevic refused to accept any terms that would allow NATO forces into Kosovo.

NATO's answer? *Air power.* First to convince, then to compel.

Their strategy rested on two assumptions: (1) That NATO could shatter the Yugoslav military without a land invasion; and (2) that the public sentiment inside Belgrade would crack before NATO's patience did. In practice, it meant a battle rhythm of all-weather airstrikes; electronic suppression; interdiction of ground forces, bridges, depots, and air defense nodes; along with constant fighter patrols to make sure any MiG-29s that flew wouldn't stay airborne for long.

Air planners stepped their way north by design. Phase I pushed hard against enemy radars, SAM batteries, and command centers; Phase II hit fielded forces below the 44th Parallel; Phase III slammed targets around Belgrade and deeper into Serbia.

Allied Force stayed airborne even as the weather turned sour, and even when errors tragically killed civilians, as in the mistaken strike on a refugee convoy in April. What didn't change, however, was the grinding math of sortie generation. Indeed, by the end of Allied Force in June 1999, hundreds of aircraft had cycled through the day and night missions; fuel tankers became a skyborne railhead; and the F-16 community (American and European) shouldered a disproportionate share of the dirty work along the enemy's biggest threat rings.

On the American side, the *Block 50* F-16CJ was the spearhead of the Allied SEAD effort. With HARMs and combat-proven HTS pods, CJ flights prowled the edges of enemy airspace, baiting SA-3 and SA-6 crews into brief, fatal exposures. But the CJs didn't work alone; EA-6Bs smeared the spectrum while RAF Tornados brought their own HARMs. But it was often the Viper that pressed closest, pinging enemy radars in the language of azimuth, timing, and intent before sounding off with an AGM-88. Throughout Allied Force, the SEAD enterprise relied heavily on the F-16CJs, and their presence was a constant hum beneath the strike packages that followed.

Within the DATF, the Royal Netherlands Air Force came in with the quiet confidence born of a Mid-Life Update (MLU). The MLU overhaul had turned their F-16AMs into all-weather, BVR killers. Their jets were among the first NATO aircraft to enter Serbian airspace from Italian air bases. From their expeditionary headquarters at Amendola, the DATF rotated their F-16s across various taskings from sweeps to strikes to bomber escorts.

Belgium's Vipers had waited a long time for this kind of war. Indeed, Belgian jets hadn't flown in combat since the Congo crisis. Allied Force changed that overnight. Twelve Belgian F-16s deployed to Italy, flying 679 combat missions between March 24 and June 10—some with MLU upgrades, others not. The distinction was important because the MLUs shaped mission planning. MLU jets tended to fly at night and tackled more demanding roles, while the non-MLU F-16s were relegated to daytime missions. The three Belgian squadrons seamlessly integrated with the Dutch at Amendola, a partnership so intertwined that, at times, the DATF felt like a single air wing with two flags sewn onto its flight suits.

But that integration paid off. Belgian pilots flew CAP and escort, delivered precision-guided bombs, and held the line in the tedious work of air policing over hostile territory. Amendola's ramp became a multilingual choreography: Dutch crew chiefs conferring with Belgian armament specialists; American liaison officers organizing threat updates; and Italian base ops who kept the day-to-day logistics running smoothly. The sortie flow was relentless, and the discipline absolute—no one needed a lecture on the consequences of an improper load or a wrong coordinate. Over the north Adriatic, aerial tankers logged hundreds of refuel missions with Belgian F-16s. In the black beyond, Yugoslavian radars tested their luck, and sometimes paid dearly for it.

The Royal Norwegian Air Force arrived with modest numbers (only six F-16s) but carried an important job: Holding CAP stations to keep the Serbian Air Force pinned. Operating from Grazzanise Air Base, Norway's jets flew long, often quiet missions, scissoring orbits while the strike packages wreaked havoc and CJs played their games of electronic cat-and-mouse. It wasn't glamorous work; and according to one Norwegian pilot, it "wasn't the war the RNoAF had trained for." But the mission mattered. Serbs knew better than to sortie a MiG-29 when an Allied fighter patrol was blocking their only means of egress.

Norway's contribution also exemplified something that Allied Force had imposed upon every NATO participant: Adaptation. In 1999, the Norwegians hadn't yet deployed their MLU jets. Their F-16s were better suited for aerial combat than the precision ground-attack roles that many NATO partners flexed. And that was fine—the coalition needed an air policing network more than it needed another half-dozen bomb droppers. The Norwegians found their niche and held it, night after night.

On the morning of March 24, 1999, the Yugoslav Air Defense Command was prepared for battle. The 204th Fighter Regiment, for example, had redeployed their MiG-29s to critically-positioned airfields throughout Serbia and Montenegro. At the same time, the Yugoslav Air Force had consolidated their MiG-21s and MiG-29s into a series of "task forces." These mixed fighter detachments were consolidated at Nis, Ponikve, and Podgorica Air Bases—each consisting of two MiG-29s and two MiG-21s.

At 7:30 PM local time, Yugoslav air defense radar detected the first wave of incoming aircraft over Albania—an American strike package consisting of EA-6Bs, F-15s, and F-16s. A few minutes later, US and Royal Navy vessels in the Adriatic Sea launched their first cruise missiles against select targets in Yugoslavia. Meanwhile, from other Italian air bases, participating NATO allies launched their own aircraft—including RAF Harriers, German

Tornados, French Mirage fighters, and the Dutch F-16s. By 8:00 PM, Slobodan Milosevic declared that Yugoslavia was now at war.

There were two waves of Allied airstrikes on the first night of the campaign. The first wave began at 7:41 PM on March 24, and lasted through midnight; the second wave launched during the pre-dawn hours of March 25, lasting from 1:00-3:30 AM. During that time, a handful of Serbian MiG-29s scrambled to meet the threat—none of which would survive their encounters with NATO aircraft.

Within the first few hours, NATO aircraft had confirmed multiple kills, mostly from the single-seat F-15s. The first F-16 victory of the operation, however, came not from an American Viper, but from the hands of a flying Dutchman. Major Peter Tankink, of the Royal Netherlands Air Force, was flying as part of a four-ship detachment from the Dutch 322nd Fighter Squadron. That night, the DATF F-16s were flying cover for the US-led strike package into Yugoslav territory. Tankink's aerial victory was the last one recorded for the night of March 24, but it mattered symbolically. The Royal Netherlands Air Force hadn't scored an air-to-air kill since the 1940s. But doing so on the first night of Allied Force, as part of a multi-national package, with a BVR shot from an MLU Viper, validated decades of investment and painstaking MLU processes. It also sent a clear message to the Yugoslav Air Force: NATO wasn't bluffing.

As Tankink's four-ship crossed into Serbian airspace, an AWACS call sliced through the radio net—three MiGs, airborne and hunting. Seconds later, his radar painted a solitary return. It was a MiG-29 flown by Major Predrag Milutinovic. Launching from the mixed fighter detachment at Ponikve Air Base, Milutinovic was the fifth Serbian pilot to scramble that night, wheels up at 8:45 PM to meet the incoming NATO strike packages head-on. But almost as soon as he leveled off, his radar faltered, another victim of the embargo-starved maintenance program and the ongoing parts cannibalization.

His radar was dead, but the threat-warning receiver still had a pulse—two sharp tones blaring in his headset, each one indicating that hostile aircraft were close. But Milutinović made the split-second choice to bypass the threats, angling instead toward Ladjevci Air Base. The field was under full blackout, but it gave him a brief window to re-establish radio comms with his squadron leaders. The order came quick: Divert to Niš Air Base, ninety miles east.

The hop should have taken no more than fifteen minutes. But as he climbed to 6,000 feet above the dark bulk of Mount Jastrebac, he was hit by a sudden and violent jolt. At first, he thought it was anti-aircraft fire raking him from below.

In reality, it was Peter Tankink's F-16.

He had just fired an AIM-120 at Milutinovic.

A mere thirty seconds later, the lurking MiG-29 disappeared from Tankink's radar screen. Milutinovic, startled by the sudden hit, promptly ejected from the stricken plane. After safely parachuting to the ground, he was recovered by some local townsfolk and taken to a nearby hospital in the city of Krusevac. Back at Amendola, Tankink's ground crew discreetly painted the tiny silhouette of a MiG-29 under the canopy rail of his F-16— a visual reminder of the aerial victory achieved by the plane and its pilot.

Although the first month of Operation Allied Force had been a resounding success for NATO air forces, the coalition did not emerge completely unscathed. For instance, on March 27, 1999, an F-117 Nighthawk from the US 8th Fighter Squadron was shot down 25 miles northwest of Belgrade. Although the aircraft bore the name "Capt. Ken Dwelle" on the fuselage, the F-117 had actually been piloted by Lieutenant Colonel Dale Zelko on the night of the mission. Luckily, Zelko ejected from the downed Nighthawk and was safely recovered by an Air Force Pararescue team.

The culprit was later identified as an SA-3 fired by the Yugoslav Army's 3rd Battalion, 250th Air Defense Missile Brigade, commanded by Colonel Zoltán Dani. Rumors circulated that the F-117 had been downed by a Serbian MiG-29, but those claims were later debunked. Still, the downing of a vaunted plane like the Stealth Fighter illustrated the perils of Yugoslavian air defense. The Serb missileers were clever: Silent radars; pop-up engagements; shoot-and-scoot batteries using mobility and the clutter of Balkan topography. For every radar shut down by HARMs, there was another that lived to blink another night. It was a high-speed chess match played at Mach 1, sometimes on instruments and often in bad weather.

And the SAM rings weren't static. They hunted back.

In Serbia, the SA-3 and SA-6 crews were old hands with years of combat experience. They understood suppression and they sought to survive it. But the F-16CJ's answer was flexibility: Feints, orbits, and vectors from opposite bearings—forcing a radar to choose its poison. Even so, sometimes the bad guys guessed right. Such was the case on the night of May 2, 1999, when Lieutenant Colonel

David Goldfein, commander of the 555th Fighter Squadron, was shot down by a Serbian SAM.

That night, Goldfein was leading his F-16 flight over western Serbia on a mission to hunt and destroy enemy air defenses. The campaign had been grinding for weeks. Aerial tankers crisscrossed the Adriatic, AWACS painted living maps in the sky, and pilots learned to read the silence between radio calls. The mission that night was simple in theory, but lethal in intent: Find the SAM batteries threatening NATO aircraft, and make them die quietly. But for David Goldfein, it turned into a raw, primitive fight against the missile he never saw coming.

The kill-chain began not with guns but with an invisible pulse. According to after-action reports (and Goldfein's own recounting), warning lights and instruments blared and the F-16 shuddered. An SA-3—reportedly fired by the 250th Air Defense Missile Brigade—detonated near the belly of his F-16. Coincidentally, it was the same unit that had shot down Dale Zelko's F-117 weeks earlier. The blast from the SA-3 didn't knock Goldfein from the sky right away. Rather, it riddled the jet's fuselage, peppering shrapnel into the cockpit. He felt the sting, later describing a minor shrapnel wound to his hand, and felt the aircraft begin to die underneath him.

Still, there was an unusual peace in Goldfein's voice as he keyed the radio. His transmission, captured on tape and retold in multiple accounts, was steady and professional.

"Start finding me, boys," he said.

His words were a terse, preemptive instruction to the rescue team that he knew would come. He fought to stretch the F-16's glide as long as possible. Then, when the jet no longer offered him a chance, he pulled the ejection handle and let his parachute take over. As Goldfein later said: "That's when your training kicks in."

The ground below was a map of shadows and danger. Goldfein's chute put him down in a field deep within hostile territory, exposed and vulnerable, a small bright target under a full moon. He hid in a ravine, doing what downed aviators do to stay alive—staying low, breathing slow, and listening for the enemy. It was the intersection of luck and craftsmanship: A good parachute, a quick egress, and an immediate plan to survive until the rescue team found him.

The recovery, as it turned out, was as vital to the story as the shootdown itself. Pararescue operators and helicopter crews staged a rescue worthy of textbook training vignettes. NATO's pararescue jumper (PJ) teams—waiting on standby alert in Bosnia—rushed across enemy lines. In accounts that later read like a manual of courage, the PJs jumped into enemy airspace and orchestrated a pickup under the worst of conditions, guided by Goldfein's short-form inputs and the AWACS's eyes above. The team fought through weather, terrain, and the threat of hostile forces to get him home. Goldfein remains forever grateful to the men who rescued him.

Goldfein has spoken about that night on multiple occasions—sometimes with the clipped professionalism of a hardened pilot, sometimes with a warm fondness for the crew that saved him. During interviews, he recalled the surprise of the second missile and the almost ritualistic sequencing of his own training steps under pressure. He remembered the full moon and how being highlighted under the lunar glow made him an easy target for enemy SAM crews. He credited his survival to the training he'd undergone, and to the comrades who had dropped everything to come get him.

Context, however, sharpens these individual moments into something larger. Goldfein and Zelko were the only two American pilots shot down during Allied Force. It was an emblematic rarity in a campaign otherwise dominated by standoff bombings and SEAD strikes. The double loss of an F-16 and F-117 underscored that even with stealth and speed, anti-aircraft systems could still bite if they were fed the right variables. It was a reminder that technology and tactics were hourly bargains, not guarantees.

For Goldfein, the night of May 2 became a defining moment. He would rise through the ranks of the Air Force and speak often about leadership under pressure, always keeping the memory of the rescue teams close at hand. He later became the 21st Chief of Staff of the Air Force, serving from 2016 until his retirement in 2020.

The last aerial battle of Allied Force occurred on May 4, 1999. That morning, the sunrise broke gray and sullen over the ramps at Aviano Air Base. In the cockpit of his *Block 50* F-16CJ, Lieutenant Colonel Michael Geczy tightened his harness and scanned his instruments with the muscle memory of a man who had been there a hundred times before. He was a seasoned veteran from the F-4 Phantom stock, having transitioned to F-16s in 1986. He was also a Fighter Weapons School graduate and a two-time combat veteran with tours in Bosnia and Iraq.

Today would be his 115th combat mission, and his seventh of Operation Allied Force.

The mission was straightforward: A dual-role SEAD and fighter sweep to guard NATO's strike packages from enemy fighters and SAM batteries. Today, he was leading a four-ship formation, each Viper bristling with two AIM-120 AMRAAMs, two AIM-9 Sidewinders, and a pair of

AGM-88 HARMs. They were to escort Strike Groups Alpha and Charlie, while another flight of F-16s would cover Strike Groups Bravo and Delta. When one group refueled, the other would rotate in.

But the enemy knew they were coming.
Central Serbia was a fortress of SAM sites, the kind that had already swatted two NATO aircraft from the sky. All told, Geczy was more worried about the SAM batteries than the Serb/Yugoslav fighters.

Meanwhile, over the horizon, another man strapped into his own cockpit—Lieutenant Colonel Milenko Pavlovic, commander of the 204th Fighter Regiment. Pavlovic hadn't been scheduled to fly that day. In fact, as the regimental commander, he wasn't supposed to be anywhere near a combat sortie. But when orders came for a MiG-29 to intercept the incoming NATO package, he pulled one of his younger pilots from the aircraft, reportedly saying: *"You're* not going to die. *I'm* going to die."

Some say it was vengeance. Valjevo, his hometown, had been bombed days before. Others believed it was guilt: He had sent too many men into hopeless fights against NATO aircraft. Either way, at 8:45 AM, his MiG-29 thundered off from Batajnica Air Base, barreling towards the oncoming storm.

For Geczy, the weather over Serbia looked ominous—low ceilings, shifting cloud banks, etc. The Dutch F-16s and RAF Tornados in Strike Group Alpha had already aborted their runs. Only the French pressed on, asking Geczy's flight for eight more minutes on station, a stretch that turned into twelve minutes. By the time the strike group headed home, the F-16s were running low on fuel.

Then the call came. AWACS had tagged an inbound

bogey.

At first, Geczy didn't hear the call, but his number four wingman did, and relayed it out over the intra-flight channel. The F-16s banked hard, climbing through 30,000 feet. Then came a second call from the AWACS: The contact was now hostile.

It was Pavlovic's MiG-29.

By now, Geczy's fuel state was critical. He knew they might have to divert to Sarajevo after the merge. But for now, he had more pressing matters: His threat receiver lit up—the MiG's radar had found him.

Geczy pitched toward the threat, both AIM-120s armed and ready. Two white-hot spears streaked from his wingtips. Below, Pavlovic climbed, his own radar painting the sky. Ground control asked if he could see the enemy.

"I see them," he replied, "but they see me too."

Moments later: "They got me."

It was the final transmission before Pavlovic's MiG disappeared from radar.

From Geczy's vantage, the first AMRAAM bloomed into a fireball. The second followed, shredding what was left of the Serbian fighter. He called it in: "Splash one, with a fireball."

On the ground, people saw the kill and cheered, thinking it was a NATO jet falling. But the celebration died as the truth sank in: The last Serbian MiG-29 to challenge NATO had just been torn from the sky. Barely a half hour later, the Yugoslav Air Force confirmed that Pavlovic had been killed in action. The news sent shock waves throughout the Yugoslav military. To that point in the war, Pavlovic was the highest-ranking Yugoslav commander killed in action; and his death represented the *sixth* MiG-29 destroyed in the air war over Kosovo. The Yugoslav Air

Force consequently suspended all fighter activity for the remainder of the conflict.

But even for Geczy, the victory was short-lived. The aero-tankers had pulled back to a safer airspace, forcing the F-16s to refuel in worsening weather and amidst a crowded queue of thirsty aircraft. By the time they limped back to Aviano in the torrential rain, the mission for Strike Group Charlie had been scrubbed.

For Pavlović, it was the end. For Geczy, it was the last air-to-air victory of Operation Allied Force.

The Yugoslav Air Force's once-proud MiG-29 fleet was already a shadow of its former self. Pilots were logging barely ten hours a year, airframes had been patched, and avionics had been decreed "serviceable" without overhaul. Radars blinked out mid-flight and radios sputtered into silence. The early days of the war had seen Serbian pilots in a desperate scramble to score a kill. Now, with Pavlovic gone, the survivors disappeared into hardened shelters, never to rise again.

Operation Allied Force ended on June 10, 1999. Per the Kumanovo Agreement and UN Resolution 1244, Yugoslav forces withdrew from Kosovo and the region was granted full autonomy. NATO had won the air war without the loss of a single pilot; Milosevic capitulated; and Kosovar refugees returned home under KFOR protection. The Yugoslav Air Force had taken a tremendous beating. By the end of the war, they had lost nearly 50 aircraft, 40 airmen killed in action, and 110 wounded. But Allied Force had also exposed the liabilities of a precision war fought under tight political constraints. Weather disrupted aerial surveillance and reconnaissance. Targets moved and hid. Collateral damage—whether a misidentified convoy or a missile that had gone astray—were amplified in primetime news.

And the enemy adapted.

Decoys, mobility, emissions control, and discipline made many Yugoslav units hard to kill. Still, the airmen of Allied Force had brought the fury of war into Yugoslavia's barracks, depots, and motor pools.

Lessons learned by the F-16 community were invaluable. First, the air war demonstrated that SEAD was a continuous process. In other words, the game of cat-and-mouse never truly ends. Wild Weasels may suppress the enemy; but the enemy soon learns to live under suppression, popping up in the seams. Just as they had done during Deliberate Force, F-16CJ sorties made the rest of the campaign possible, but never truly "safe."

The Dutch and Belgian F-16s, meanwhile, learned the critical value of their MLU updates. Indeed, the upgraded radars, BVR missiles, and night/all-weather avionics meant they could shoulder more complex tasks and do it around the clock. It's not that non-MLU jets were useless; far from it. But the coalition's most flexible taskings tended to befall jets with the best weapons and sensors. Holland's own Ministry of Defense has acknowledged how useful the MLU jets were to the mission.

Most of all, the joint nature of the mission underscored how coalitions amplify combat power. Norwegian CAPs, for example, freed American fighters for strike missions; Dutch and Belgian F-16s were a "force multiplier" where it mattered the most; and the Wild Weasel F-16s defanged the enemy's air defenses so others could live. The whole of the coalition wasn't just *greater* than the sum; it was *different*—a living thing that could flex when one nation's rules were tighter or looser than another's.

After the war, Yugoslavia's air force continued along its downward descent. Flight hours and readiness rates plummeted further due to the ongoing supply shortage. In fact, the Yugoslav Air Force ceased to exist in 2003. When the post-Milosevic government dropped its claim of successorship to the Tito era, the "Federal Republic of Yugoslavia" voted itself into the latter-day confederation of "Serbia and Montenegro." In keeping with its new identity, the Yugoslav Air Force then became the Air Force of Serbia and Montenegro.

The name change, however, did nothing to alleviate the ongoing readiness issues. By 2004, reports had surfaced that Serbia and Montenegro had grounded its entire fleet of MiG-29s, as they no longer had the means to keep them flying. Two years later, Montenegro seceded from the union with Serbia, thus leaving the two entities fully independent, and formally dissolving the last remnants of the Yugoslavian identity.

From the quiet patrols of Operation Sky Monitor to the thunderclap finale of Operation Allied Force, the F-16 carved a relentless presence in the skies over the Balkans. What began as watchful orbits soon escalated into a grinding air war that tested the Viper and its pilots across every role in NATO's arsenal: CAP, SEAD, CAS, strike escort, and air superiority.

Sky Monitor was meant to be a passive, simple enforcement of no-fly zones after Bosnia disintegrated into war. But for the Viper pilots orbiting high above, it was a waiting game under tense skies. Their cockpits became crucibles for the human psyche—one eye glued to the threat receiver, the other eye fixed on the mountains below, where villages burned and paramilitary forces clashed.

When Operation Deny Flight replaced Sky Monitor in 1993, the rules changed. The F-16s were no longer witnesses; they were enforcers. As the skies over Bosnia grew lethal, American Vipers tangled with Serbian jets for the first air-to-air kills in NATO history. It was a violent baptism by fire for an alliance still trying to find its footing in a post-Cold War era.

By the summer of 1995, the fragile experiment of "containment" had given way to the fury of NATO firepower. Operation Deliberate Force saw the Viper's most intense combat mission since Desert Storm. Squadrons from Aviano and Spangdahlem flew hundreds of sorties into some of the most heavily defended airspace in Europe since World War II. In many ways, the F-16 became NATO's airborne scalpel, delivering precision strikes to ammunition depots, airfields, and command posts while braving the threat of hostile SAMs.

Then came the Kosovo War, wherein Operation Allied Force was the crucible. F-16s carried the fight from the opening salvo to the war's final moments. The *Block 50* F-16CJs hunted radars with deadly HARMs, diving into Serbia's lethal SAM belt. Their *Block 40* counterparts delivered laser-guided bombs against bridges, airfields, and hardened bunkers. And when Serbia threw its MiG-29s into the fray, Viper pilots like Peter Tankink and Michael Geczy intercepted them in short-order dogfights. The engagements were brief, brutal, and very much one-sided. NATO training, tactics, and avionics overwhelmed the brave (but under-resourced) Serbian pilots who were flying MiG-29s long past their prime. Yet, every time a MiG lifted off, the tension spiked. Every NATO pilot knew that a Serbian R-60 missile could just as easily find him.

In many ways, the Balkans had been a long-term proving ground for the F-16 and its crews. It didn't have the glamor of Desert Storm, but it carried something harder—a long-term battle of endurance, flown over rugged terrain, in an overly-restrictive political climate, and against a stubborn enemy who adapted at every turn. For the Viper pilots who took to the skies between 1992-99, the Balkans weren't merely a sideshow; they were the main event of the post-Cold War era. It was *the* battlefield where NATO tested its will and the F-16 cemented its reputation as the workhorse fighter of its generation.

In the end, the Viper flew not just as a machine of war, but as a symbol of international resolve. In skies darkened by politics and hatred, the roar of its engine sent a clear message to allies and adversaries alike: NATO was watching, and the F-16 wouldn't leave until the war was over.

Serbian MiG-29. During the Kosovo War, the MiG-29 was the primary air-to-air threat to NATO fighters. However, logistical and readiness issues hamstrung their effectiveness. Of the six MiG-29s that were shot down by enemy fire, two were downed by F-16s (one Dutch; one American). *Krasimir Grozev*

J-21 Jastreb on display at Ladevci Air Base in Serbia. This aircraft is similar to those shot down during the 1994 Banja Luka Incident by American F-16 pilot Rob Wright. Coincidentally, Wright's longtime wingman was Scott O'Grady, and the two were flying together when the latter was shot down in July 1995. *Srdan Popovic*

An F-16CJ (the dedicated SEAD variant of the Fighting Falcon) from the 52nd Fighter Wing based at Spandahlem Air Base, Germany, breaks away from a KC-135 aerial tanker after in-flight refueling during Operation Allied Force on March 31, 1999. *US Air Force*

Chapter 5:
This Kind of War

The morning of September 11, 2001, permanently split modern air power into an era of *before* and *after*. When the first hijacked airliner struck the North Tower of the World Trade Center, the United States was thrust into a new kind of war—a conflict without borders, against enemies who had no capital city to defend, no industrial heartland to bomb, and no conventional forces to eliminate. Within hours, the US had identified Al-Qaeda as the architect of the attacks, and the Taliban regime in Afghanistan as its protector. The decision was swift: Eliminate Al-Qaeda's leadership and their safe havens abroad.

But the forthcoming conflict would be nothing like Desert Storm, or even the politically-restrictive "peacekeeping" missions of the 1990s. This war would be something harsher, murkier, and infinitely more personal.

Afghanistan had been a graveyard of empires long before the first F-16s touched down at Bagram Air Base. The Taliban had risen to power in 1996, emerging from the wreckage of the Afghan Civil War, and had since turned the ailing republic into a brutal Islamic theocracy. That same year, they offered sanctuary to Osama bin Laden (the founder of al-Qaeda) whose terror network stitched together jihadists from across the globe. Thus, Afghanistan had not only become a failed state, it was now a launchpad for global terrorism.

Bin Laden and his acolytes found sanctuary among the Mujahideen remnants—training camps nestled in the ragged columns of the Hindu Kush. Airfields and former

Soviet barracks now doubled as safe havens, and the Taliban's protection had allowed al-Qaeda to perfect its murderous craft of terrorism. The World Trade Center, the Pentagon, and the charred field in Somerset County, Pennsylvania had all been conceived, planned, and greenlit from Afghan soil.

The American-led response—codenamed Operation Enduring Freedom—would not be a "conventional" war, but a campaign to destroy the asymmetric threat that had embedded itself into the mountains of Afghanistan.

In the weeks following September 11, the Taliban ignored President George W. Bush's ultimatum to deliver Osama bin Laden or face destruction. On October 7, 2001, the US military replied with the fury of decisive air power. American bombers, cruise missiles, and fast jets began carving contrails across the Afghan skies.

But this was no set-piece air campaign in the style of Desert Storm. Targets were fleeting, Taliban units melted into villages, and caves swallowed enemy ordnance whole. What NATO brought to bear was speed, persistence, and a level of precision that could thread bombs through the narrowest cave or a bunker window. But the battlefield itself seemed to resist control, daring the Western coalition to impose order on a terrain that had humbled foreign armies for centuries.

For F-16 pilots, Afghanistan was unlike any war they had ever flown. In Iraq, the Viper had faced radar-guided missiles, anti-aircraft guns, and the well-defined signatures of Soviet-style divisions. In Serbia, they had hunted mobile SAMs and dodged heavily-layered air defense rings. Afghanistan, by contrast, was both barren and invisible. No sprawling air defense systems waited to light up the sky. No fighter units scrambled to meet NATO over the mountains. Indeed, this enemy threat was different: AK-47s rattling at patrols, mortars lobbed from village walls, and RPGs streaking upwards in desperate arcs. For an F-16 pilot, "danger" wasn't an incoming MiG. It was the possibility of dropping bombs too close to an American patrol, or mistaking a farmer's field for an enemy redoubt.

The environment itself was as formidable as any foe. Afghanistan's mountains were among the tallest in the world, with peaks scraping above 20,000 feet, and narrowly-threaded valleys where insurgents could vanish in seconds. The thin, mountainous air could rob engines of thrust and stretch weapons envelopes to their limits. Pilots found themselves pulling Gs at altitudes where wings felt sluggish and climbs strained against gravity. Dust storms swept across entire provinces, grounding flights and/or blinding sensors. Even at their sprawling bases—including Bagram, Kandahar, and the outposts carved from old Soviet airfields—NATO airmen fought the elements as much as the Taliban. These included brownout landings, ice formations that locked canopies shut, and summer heat waves that warped metal.

And then there was the tempo. Unlike Desert Storm's sequential, choreographed ATOs, Afghanistan was built around immediacy. A Special Forces team in the hills might take fire and, within minutes, the call would echo through the radio net:

"Troops in Contact!"

Pilots who thought they were in for a routine armed overwatch suddenly found themselves rolling in hot, guided by the steady voice of a Joint Terminal Attack Controller (JTAC) on the ground. There was rarely time to study a target. Every mission became an exercise in trust, precision, and restraint.

This was the crucible where the F-16 earned its Afghan reputation. Gone were the dogfights and missile duels that it had been designed to tackle. Instead, the Viper's task was both narrower and far heavier: To be a proverbial "scalpel" and "sledgehammer," delivering ordnance so close to friendly troops that a margin of error measured in mere feet could mean life or death. Pilots learned to orbit for hours, sensors pinging the ground below, hunting for the telltale sign of an RPG team or the faint signature of a truck moving under the cover of darkness. The cockpit became a surveillance post, a lifeline, and a trigger point for immediate steel and fire.

In Afghanistan, the F-16 wasn't simply a fighter. It was a guardian angel occupying a 20,000-foot orbit, a constant reminder to the Taliban and al-Qaeda that no firefight was ever theirs alone. But to the men and women flying those missions, it was also a reminder that the rules of war had shifted. This was not about overwhelming an enemy with concentrated air power. It was about patience, precision, and the relentless grind of protecting soldiers in a war where the frontlines were everywhere and nowhere all at once.

Opening Rounds (2001-02)

By the time the first F-16s began orbiting Afghanistan in late 2001, the mission sets had already diverged from anything the Viper community had known in Desert Storm or the Balkans. The jet was still the same: Lean, agile, and designed for cutting-edge dogfights. But here, the fight was about presence and timing.

The most common sortie flown by the F-16 was armed overwatch. A pair of Vipers would launch from bases at the edge of the Afghan theater—first from Al Udeid in Qatar; later from Bagram and Kandahar—and climb into the thin, frigid air above the Hindu Kush mountains. Once on station, their role was deceptively simple: Watch, wait, and be ready to strike. But this kind of overhead support required immense discipline. For hours at a time, pilots flew racetrack patterns over the Afghan wilderness, their eyes shifting from the horizon to their targeting pod screens—scanning villages, roads, and tree lines. On the ground, patrols of US Army, Marine, and Special Operations forces pushed deep into hostile valleys.

Above them, the F-16 became a constant, invisible shadow. Every pilot knew the radio call that could change everything. Whenever a JTAC or Special Forces ALO keyed the radio with the phrase, "Troops in Contact!", the atmosphere in the cockpit shifted instantly. The chatter died, fuel consumption was recalculated on the fly, and the wingmen tightened formation. The request usually came in clipped and urgent: An enemy ambush, RPGs and AK fire pinning down a squad, with mortars thundering in from the hillsides.

In these moments, the Viper was no longer a patrol aircraft. It became a strike fighter in the purest sense. Pilots descended through the layers, maneuvering through valleys where airspeed bled quickly in the thin mountain air. With JTACs marking targets by laser or smoke, the F-16s would roll in, weapons armed. These 21st Century Vipers carried an array of cutting-edge armaments: GBU-12 laser-guided bombs, JDAMs for GPS-level precision, some cluster munitions, and the 20mm Vulcan as a last resort.

But Afghanistan also required finesse. The Taliban rarely massed forces or exposed columns that begged for destruction. Instead, they struck from compounds, tree groves, or rocky outcroppings, often within sight of civilians. CAS operations thus had to be executed with a surgical precision. One pass might drop a 500-pound bomb on a ridgeline, the next might require a low, intimidating "show of force"—the F-16 diving low over a village at supersonic speed, the roar being enough to scatter insurgents without firing a shot. Pilots learned that sometimes the most powerful weapon wasn't ordnance, but the mere psychological presence of the Viper screaming overhead.

Night sorties, however, added another layer of complexity. Equipped with LANTIRN targeting pods, F-16s scanned for insurgent movement via infrared, their sensors able to spot the faint signature of motorcycle headlights from as far as ten miles away. UAV drones often cued the F-16s onto suspicious movement, but the responsibility of positive ID always rested on the pilot. Under the cockpit glow, their gun reticles silently tracked shadowy figures in the night.

Were they farmers?

Smugglers?

Taliban fighters emplacing an IED?

One wrong call could permanently disrupt the loyalty of a nearby village.

Endurance was another enemy. Afghanistan's remoteness meant long transit flights, sometimes over Iran or Pakistan, flown under strict diplomatic clearance windows. A single sortie could stretch eight to ten hours, requiring multiple refuels from the in-theater KC-135s or KC-10s. In fact, refueling became ritual unto itself: Slipping behind the boom at night, turbulence rocking the Viper as it gulped down fuel, with the pilot's mind never fully disengaging from the JTAC calls murmuring in his headset.

On paper, the F-16 mission profiles over Afghanistan looked repetitive: Orbit, respond, refuel, then orbit again. But in practice, every sortie was its own crucible. Some ended with hours of uneventful overwatch. Others climaxed in split-second decisions—whether to drop a bomb inside 200 meters of friendly forces, whether to fire cannon rounds with civilians nearby, or whether to abort the mission because fuel was low and the tankers had been recalled for poor weather.

But this tempo bred a new kind of combat aviator.
The young Viper pilot, once trained to think in terms of kill ratios and missile envelopes, now lived in a war defined by restraint and responsibility. Each mission blurred the line between fighter jock and guardian angel. They became the long-term watchmen in the sky; their jets transformed from Cold War dogfighters into lifelines for men fighting inch-by-inch against the savagery of jihadism.

The air war in Afghanistan began with a thunderclap: Tomahawks barreling through the nighttime sky and B-2 Stealth Bombers flying halfway across the world to strike at the heart of Taliban defenses. Yet for all the shock and awe of the opening salvo, sustaining air power over a landlocked battlefield in Central Asia demanded something more: A multi-role fighter that could carry the daily burdens of tactical-level sorties.

Into that crucible stepped the F-16, carrying callsigns of the Air Force Reserve, Air National Guard, and active-duty wings who would write the opening chapters of the Viper's long history in the Global War on Terror.

The 466th Fighter Squadron (an Air Force Reserve unit

from Hill Air Force Base, Utah) was among the first to send its pilots into Afghanistan. In late 2001, they were the first Vipers to fly combat sorties in Operation Enduring Freedom, dispatched from forward-stationed air bases in the Persian Gulf. For years, many of these pilots had flown "routine" air patrols over Iraq, enforcing Operation Southern Watch. Now, they were the opening act of a new war, hurtling through the same mountain passes where Soviet MiGs had once been ambushed, carrying precision ordnance meant for Al-Qaeda and their Taliban hosts.

It was a strange twist of fate: A Reserve outfit, long accustomed to balancing civilian jobs with military duty, was now breaking ground for the entire F-16 community in Afghanistan. They were the tip of the spear in a fight that promised no clear front lines and no conventional enemy.

Flying from their bases in the Gulf, F-16s from the 466th shouldered the burden of early CAS and strike sorties, slipping into Afghan airspace where the terrain was as lethal as the enemy below. For a reserve squadron accustomed to stateside training sorties, the transition to live combat over the Hindu Kush was jarring but decisive. Every bomb they dropped, every show of force they executed, became part of the fragile lifeline that kept Special Forces teams and Northern Alliance fighters alive in firefights against Taliban strongholds.

On the second night of the war, October 8, 2001, Major Kurt Gallegos of the 466th found himself leading one of the most audacious sorties yet attempted over Afghanistan. He was the flight lead of a four-ship formation (two F-16s and two F-15E Strike Eagles) cutting north out of Kuwait into the dark, bound for a battlefield no American fighter pilot had ever seen.

Just weeks earlier, Gallegos and his squadron mates had expected nothing more than another deployment to Iraq's Southern No-Fly Zone. But after September 11, the mission changed overnight. Iraq was no longer the focus. Al-Qaeda and the Taliban had become the priority targets.

The first wave of American jets had thundered into Afghanistan the night before. Their plan had been to hit the enemy with *eight* jets at a time: Four Strike Eagles and four Vipers. It was bold, but the math didn't work. Tankers couldn't feed that many fighters, not over such long distances. Nearly every aircraft ended up running on fumes. The lesson was hard but clear: Fewer jets, longer legs. "So, the next night, they paired it down to two F-16s and two Strike Eagles," said Gallegos, "and that was a four-ship that I was the mission commander on."

The mission that unfolded would become a test of endurance: 10.8 hours in the cockpit, the longest combat sortie an American F-16 had ever flown up to that point.

Their route was a marathon across three countries. They launched from Kuwait, vectoring south over the Persian Gulf for the first refueling. In the dark, the Vipers slid up behind the tankers, drawing fuel before banking northeast across the Arabian Sea. Over Pakistan, they met a second tanker, topping off again before committing to the Afghan frontier. Ahead lay the mountainous landscape of Taliban country, dark ridgelines hiding unknown threats, and valleys where the Soviet Army had bled two decades earlier.

Once across the border, the formation split. The Strike Eagles veered off to prosecute their own targets. Gallegos and his wingman pushed deeper into the Afghan interior, their radios alive with a call from the ground. A small team of US Special Forces (Green Berets working with Afghan allies) was in contact with Taliban fighters. The commandos needed bombs on target, and fast.

"We talked to the controllers on the ground," Gallegos later recalled. "JTAC Special Forces guys…they happened to encounter a Taliban encampment." The coordinates came quickly. The Taliban camp was larger than expected, a sprawl of tents, vehicles, and fighting positions hidden in the rocks.

Gallegos rolled in.

One by one, the F-16s dropped their loads, four laser-guided bombs each. The hills flashed white as explosions tore through the encampment. Secondary fires lit up the surrounding terrain. Radios then crackled with confirmation from the Special Forces team: The strikes had decisively gutted the Taliban position. "We destroyed quite a bit of Taliban that night," Gallegos said simply.

Indeed they had.

A huge encampment had just gone up in flames.
When the last bomb fell, the two Vipers pulled off the target, climbing back into the dark beyond. They rejoined the F-15s at a pre-arranged rally point, fuel gauges once again dipping low, and turned south for the long ride home. Hours later, they touched down triumphantly onto the Kuwaiti airstrip.

It was a glimpse of what the War in Afghanistan would come to demand: Endurance flights over hostile mountains, precision strikes called in by men on the ground, and small formations carrying the burden of decisive firepower. For Gallegos and his fellow pilots, it was just the second night of the campaign. For the Taliban, however, it was the beginning of a long war fought under the fury of Allied air power.

Citizen-airmen from the 466th Fighter Squadron had paved the way for Viper operations in Afghanistan, but the active-duty squadrons would not be far behind. In mid-October 2001, the runway lights at Al Udeid Air Base were still a rumor when the first jets and support crews from the 366th Fighter Wing clawed their way into Qatar. It was a lone strip of concrete amidst a sea of brown dust—no dorms, few generators, and barely enough shade. This was the forward edge of a new kind of air war.

Engineers and maintainers slept in packed metal huts, then went straight to building the base they needed: Fuel bladders first, operations center next, tents last. Within days, aerial tankers were launching, the tent city began to rise, and the 366th started turning the desert into a tactical air hub aimed squarely at the Taliban's heart.

By early November, the 389th Fighter Squadron—the Wing's primary fighter unit, still flying F-16CJs—rolled in from Mountain Home AFB, Idaho and took its place on the chalkline. They were, as the 366th Wing Historian later noted, the first US fighter unit to operate out of Al Udeid. And, in short order, they would accrue a number of combat "firsts." One day after their arrival, the squadron commander himself was airborne on the unit's first combat sortie. Over the coming weeks, the 389th would fly a total of 244 missions, log more than 1,500 combat hours, and drop 185,000 pounds of ordnance—numbers that meant little on paper, but everything to Special Forces teams and Northern Alliance fighters pinned down in the Afghan valleys.

Al Udeid itself was the picture of improvisation. The 366th's initial cadre made fuel and Command & Control their first priorities; living conditions could wait. Six days after they hit the ground, the field went operational. The

KC-10s began cycling into the Afghan sky. It was expeditionary warfare doctrine come to life: Build forward air power where none exists, then keep it fed. In those first 105 days, the 366th Fighter Wing deployed more people and aircraft (and dropped more bombs) than any other Air Force unit, a sprint at the start of a marathon.

From Al Udeid, the 389th's F-16s launched into the kind of missions that didn't look like Desert Storm or Southern Watch. They were long ribbons of time: 8-10 hours aloft, stepping through tanker tracks, sliding across maritime airspace, and then over jagged mountains. In November 2001, Captain Craig Marion led one of those endurance runs, an eight-hour mission that bent itself around a single purpose: Keeping an Army convoy alive. There was no glamour to it—just a drumbeat of check-ins, fueling, sensor work, and the steady threat of a sudden call: "Troops in Contact!" Marion's performance in that mission earned him the 2001 Semper Viper Award, an annual award granted by Lockheed Martin to a pilot who demonstrates exceptional devotion to duty. But the essence of Marion's mission was far simpler. When the convoy called, an F-16 answered, and stayed until the danger lifted.

Even the airplanes themselves told a story. The squadron arrived with their *Block 52* CJs—SEAD thoroughbreds that, by 2001, had matured into true multirole fighters. Once in theater, the squadron borrowed a few additional Falcons from the citizen-airmen of South Carolina's 157th Fighter Squadron to round out the line, molding the Active-Guard mix that would become a hallmark of Enduring Freedom. Nose art, a time-honored tradition from the days of World War II, bloomed in the dusty ambience of Al Udeid. The iconic phrase *"We'll Take*

It From Here" stood emblazoned upon the F-16 tail-numbered 91-0401. It was a symbolic promise to the FDNY and other first responders who had given their lives on September 11.

On November 29, the 389th achieved a tactical milestone that, unwittingly or not, shaped the vocabulary of Operation Enduring Freedom. Their F-16s became the first of their kind to drop the 2,000-pound GBU-31 Joint Direct Attack Munition (JDAM) in combat. For the F-16 community, the JDAM's combat debut was a pivotal moment: GPS-guided precision meant targets didn't need to glow under a laser in clear air. Targets could now be struck through low-visibility conditions and in the dead of night. In a combat theater where weather and terrain could hide enemies as effectively as man-made structures, the JDAM's technology mattered.

But if JDAMs were the headline, the F-16's daily grind was the subtext. Missions spanned the spectrum multirole capabilities: "Push CAS" with Special Forces teams threading through the north; armed overwatch for convoys and air assault packages; interdiction strikes against vehicle columns and weapons caches; and on-call "show of force" runs to keep Taliban units moving at night instead of mounting daylight defenses. For the 389th, and other follow-on Viper squadrons, Enduring Freedom was a ground-attack war flown at expeditionary distances, from a bare-bones Qatari air base, sustained by a tanker network straddling the upper edges of the Indian Ocean.

The 389th's pace mirrored the broader tempo of its parent air wing. While their sister squadron, the 391st was carving trenches in the air over Afghanistan with GBU-12s, the 389th kept building a record of monumental "firsts" that the 366th's Wing Historian would later tally. This included the first operational uses of both JDAMs and the Wind-Corrected Munitions Dispenser (CBU-103) by an F-16, and the first F-16CJ squadron to fly CAS for Special Forces in contact. Each milestone highlighted how the air war in Afghanistan had shifted from mass bombing runs to time-sensitive, tactical air support woven into the ground commanders' scheme of maneuver.

From the flightline to the cockpit, none of the work was glamorous. The workday began with crew chiefs scribbling new chalk marks under intake lips—sorties counted, weapons expended, and jets serviced on a quick turnaround for the next mission. Over Afghanistan, Viper pilots listened intently for any sign of trouble in the cadence of a JTAC's voice; circling patiently while ground forces moved with deliberate intent. When the call came, the response was clean and fast: Talk-on, reset the master arm, check geometry, and clear hot. Then fly back to the orbit, back to the tanker, and back into the stack…until it was time to vector home to Al Udeid.

Throughout December and into January 2002, the 389th stayed on task, prying open valleys and roads for the Northern Alliance's advance and coalition raids. Their sortie counts climbed and their footprints on the map thickened wherever contact calls were hottest.

By month's end, many of the 389th's elements began redeploying home. But the austere airfield they had built into a fully-functional forward base kept humming for the remainder of the war. What they left behind was more than a tally of enemy losses. It was a *playbook*: How to build an expeditionary air hub in record time; how to prosecute fighter operations over extreme distances; and how to bend precision-guided weapons and aerial logistics to the service of small ground teams in big mountains.

In a war that would drag on for twenty years, the 389th's short, sharp deployment helped define what the F-16 would do in the skies over Afghanistan—overwatch, strike, reassure, and deter.

Anaconda **and beyond (2002-05)**

By January 2002, the 389th was gradually being replaced by the same citizen-airmen from whom they had borrowed a handful of Vipers the previous fall—the 157th Fighter Squadron of the South Carolina Air National Guard. Mobilized during the aftermath of 9/11, the renowned "Swamp Foxes" had already earned a distinguished combat history, having flown in Desert Storm and over the Southern No-Fly Zone. Beginning on January 8, the unit consolidated their assets onto Al Udeid Air Base, just as the war shifted into a higher gear. For the next three months, the Swamp Fox Vipers roared along the same aerial trails that the 389th had blazed into Afghanistan. And the Swamp Foxes' arrival could not have been timelier. For in March 2002, American troops were locked into Operation Anaconda, a ferocious battle against al-Qaeda in the Shah-i-Kot Valley.

Overhead, the F-16s of the 157th orbited as silent guardians. When troops in contact called for help, the Swamp Foxes dove in through the thin mountain air, delivering smart bombs onto targets only meters from friendly positions. The missions were high-wire acts of coordination—FACs on the ground whispering target coordinates under fire, pilots making split-second calculations as they lined up their runs, and the omnipresent risk of collateral damage to the nearby villages. Yet sortie after sortie, the Swamp Foxes proved that a National Guard squadron, mobilized from small-town South Carolina, could perform with the precision and ferocity of any Active Duty outfit.

And nowhere was their tenacity and professionalism better exemplified than during Operation Anaconda. Beginning on March 2, 2002, the rugged terrain and jagged peaks of the Shah-i-Kot Valley erupted with the sound of mortars, small arms fire, and airborne ordnance delivered at "danger close" proximities.

By all estimates, Anaconda wasn't supposed to have been this loud.

Coalition planners had imagined the Taliban and al-Qaeda scattering under pressure from Afghan allies, with US infantry sealing exits, while precision air power finished the job from above.

Instead, the valley bit back.

Enemy fighters dug into the high ground with heavy machine guns and mortars trained on the likeliest avenues of approach. Throughout the ensuing battle, helicopters took hits and assault elements were pinned down. Suddenly, what should have been a "shaping" operation became an all-out fight for survival, with F-16 air support surging from their orbits to rugged terrain below.

By March 2002, the F-16s were no strangers to Afghanistan. Day after day, they flew long, multi-hour sorties from their bases in the Persian Gulf, armed with JDAMs or GBU-12s, proving to be the nimblest and hardest-hitting tactical air assets in theater. Whenever platoons on the valley floor called "Troops in Contact," it was often a pair of F-16s that would drop quickly from mid-stack, get eyes on the target, and deliver their bombs with devastating accuracy under the control of JTACs, Air

Force Tactical Air Control Parties (TACPs), and attached Air Liaison Officers (ALOs).

The opening days of Anaconda revealed just how hard this valley would be. Communications were brittle. Some Army elements lacked direct, robust digital links to the COAC. Meanwhile, ALOs and TACPs were stretched thin across dispersed units; and the sheer number of aircraft meant the stack sometimes looked like a waiting room during a mass-casualty event. Air controllers fought to keep flows deconflicted while prioritizing targets that changed minute to minute—enemy mortars during one pass, machine-gun nests on the next pass, then a call to clear a landing zone for a battered Chinook. During all that chaos, the F-16 crews fell back on their disciplined CAS habits: Check in, authenticate, get the 9-line, confirm friendlies and geometry, and go.

On the ground with the 101st Airborne Division stood Captain Paul Murray, an Air Force F-16 pilot serving as an attached ALO. Murray's job in this battle wasn't to pull high G turns; it was to make air power talk to soldiers under fire. He moved with the division's 3rd Brigade as it fought for the heights, working radios and terrain to paint a picture for the fast-movers overhead. Murray and the TACPs around him served as the tactical "translators"—turning a squad leader's shouted contact report into coordinates that an F-16's pilot could find, and line up a trajectory that wouldn't turn the GBU-12 into an instrument of fratricide. Accounts from official studies and Murray's own after-action reviews underscore how ALOs (like him) became critical nodes once the battle plans collided with reality. He worked from the dust and rocks (not from the confines of the CAOC) and when an F-16 checked in overhead, the pilot often found Murray's voice on the net.

Throughout Anaconda, call-for-fire sequences typically followed the same pattern: An Army element travelling along a spur would get bracketed by mortar fire, then get strafed by DShK and/or RPG bursts from a saddle they couldn't reach. Murray or a TACP would then key the radio, and pull a pair of F-16s off the stack with a direct check-in. While the flight leader slid his targeting pod across the ridge, the ALO confirmed the friendlies' exact position—bearing and distance from a reference point, laser code if available, and any No-Fire lines. Then came the rapid-fire 9-line. The F-16 wingman would set for a Type 1 or Type 2 control depending on visibility and proximity, then roll into a precise delivery: GBU-12 against a machine-gun nest dug into a loophole, or a JDAM onto a mortar pit revealed by muzzle flashes registering on the pod's IR sensor. The first pass might miss the last rock by mere inches; but the second would walk the coordinates into the trench. On more than one occasion, the sound of a pair of Vipers crossing the bowl, and the brief thunder of their bombs, spelled the difference between life and death.

The battle also forced some hard lessons across the joint team. Early frictions—including airspace control, communication gaps between the Army and Air Force operators, and a scarce number of JTACs to cover multiple firefights simultaneously—occurred in real time. But the air and ground assets adapted together. The CAOC reshuffled their priorities on the fly. Stacks were re-tiered to shorten their time-on-target. And with ALOs like Paul Murray inside the Army's scheme of maneuver, coordination got tighter by the day. Post-operation analyses highlighted these shortcomings and the fixes they drove afterward: New liaison parameters, better ASOC processes, and a wider distribution of trained JTACs. But in those heated moments of March 2002, it was the people in the loop who made it work.

There was no single "F-16 moment" that defined Operation Anaconda. Indeed, there were dozens: A pair of Vipers clearing a landing zone for a medevac arrival; a section rolling in danger-close to kill an RPG team in contact; another pair working with an AC-130 to obliterate the enemy on a contested ridgeline. Across the valley, the F-16s' contribution was precision and responsiveness—jobs that required fewer bombs than the B-52s, but more coordination than any single aircraft could do alone. The result, by mid-March, was a fire-marked valley that the enemy could no longer hold. Coalition ground forces had pried them from the high ground of the Shah-i-Kot Valley, and air power (jets and gunships alike) had been decisive at every turn.

Murray's role was a reminder that CAS is as much a relationship as it is a weapon. Murray wasn't just a voice on the headset; he was an F-16 pilot who knew how Viper crews would see a ridgeline through a pod. He knew what information they needed to feel confident on a Type 1 or Type 2 control. And he knew how to trim a 9-line to what mattered when rounds were hitting close.

He also knew what a platoon leader under fire needed to hear from an aircraft overhead: Confirmation that the target was the right one; a time-to-impact that was honest; and a willingness to do it again if the first pass didn't solve the problem. His performance, and that of the TACPs spread thin across the battlefield, shows up in the sobering language of official histories and after-action interviews. But the effect was visceral: Each controlled strike took the pressure off American soldiers, and put it on the jihadists who thought the valley was theirs.

One notable F-16 mission from the perilous days of Anaconda occurred on March 4, 2002. For Lieutenant Colonel Burt Bartley, commander of the 18th Fighter Squadron, that morning began like many others during Operation Anaconda—a stacked orbit crisscrossed with bombers, surveillance aircraft, tactical fighters, and limited ordnance. His squadron had deployed from Eielson AFB, Alaska to Ahmad al-Jaber Air Base, Kuwait in December 2001. And, like the 389th and 157th, his F-16s had taken the mantle of tactical air support. While orbiting the battlefield from above, an urgent call broke in: An MH-47 Chinook had been shot down, its crew pinned under an emerging wave of enemy fire. The primary airborne assets—two F-15E Strike Eagles—were already on station, but they had expended their ordnance. The valley was closing in, and time was running out for the thirty-nine American troops pinned in a deadly ambush.

Hearing the urgency over the radio, Bartley immediately broke from his assigned airspace. He knew what was at stake; every second counted. On his way in, he carefully deconflicted with two UAVs and the Strike Eagles, making sure the airspace was safe so his F-16 wouldn't clash with another asset.

Closing in on the rescue site, Bartley didn't hesitate. He made two strafing passes, pulling low through rising terrain, diving into the face of the same hostile fire that brought down the MH-47. He unleashed 500 rounds from the barrels of his 20mm autocannon, pummeling enemy positions and forcing them to break contact a mere 50 meters from the crash site.

But the fight wasn't over. His wingman, now having radio issues, couldn't relay updates to COAC or the other airborne assets. Without a moment's pause, Bartley took control of the comms, directing his wingman to engage advancing enemy forces as he coordinated with rescue elements and directed a nearby tanker to position for further support.

When the enemy tried to reassert, the FAC called Bartley to drop his 500-pound bombs to within 100 meters of the crash site—close enough to be dangerous, but deliberate enough to be effective. Bartley complied, shaking the ridge and sending the enemy into a hasty retreat. With the area now contained, the rescue helicopters swooped in. Bartley valiantly guided them to the crash site, escorting the choppers through a corridor still teeming with small-arms and RPG fire, giving them a clear path to reach all 39 survivors. Every one of them—injured or intact—made it out of the kill zone.

For his actions that day, Burt Bartley received the Silver Star—a testament to his quick judgment, fearless valor, and tactical airmanship under fire. For the F-16 community, it showed what a Fighting Falcon and its pilot could still do: Rapid-fire force protection and precise target engagement in a high-stakes gunfight.

In many ways, Operation Anaconda became a benchmark. It showed what precision air power could accomplish inside a rugged, high-altitude battlespace when it was tied directly to ground maneuver. For the F-16 community, it reaffirmed the jet's ongoing relevance to CAS missions. For the tactical air-ground team, it accelerated reforms that would echo throughout the rest of the War in Afghanistan. None of it was neat. Much of it was improvised. But the narrative was clear: When the Shah-i-Kot valley tightened around American units, the F-16s brought salvation in the form of some well-placed

JDAMs.

But the CAS mission demanded speed. And speed demanded proximity. The F-16s needed a base at the enemy's doorstep. And they found it in Bagram, a husk of an airbase, abandoned since 1989. Once a decaying monument to past empires, it was about to be reclaimed as the hub for Allied air operations.

When the first American planes touched down at Bagram Air Base in 2001, it was more a "reclamation" than a true arrival. The Soviets had once used Bagram as the primary hub for air operations during the Afghan War; and when they withdrew in 1989, the base had been left to rot in the crossfire of the Afghan Civil War. What remained by the time American forces arrived was little more than a scarred airstrip and the skeletal remains of hangars pocked by shellfire. Minefields surrounded the perimeter, and the control tower was little more than a gutted husk. It was hardly the kind of place one would expect to stage the world's most advanced tactical aircraft. But necessity had a way of bending expectations.

The first crews who deployed into Bagram found themselves more construction workers than combat airmen. Engineers bulldozed unexploded ordnance off the runways. Pilots and maintainers pitched tents where hangars had collapsed, stringing power lines and heating units through the bitter Afghan fall and winter. Maintenance stands were improvised from salvaged Soviet equipment. In place of hardened aircraft shelters, aircraft were parked in the open, their tails jutting skyward against the backdrop of the Hindu Kush. The constant thud of distant mortars, artillery, and the occasional crack of small-arms fire from beyond the gate reminded everyone that Bagram was not yet "secure."

Still, the jets flew. From the moment the first Allied aircraft taxied onto the fractured tarmac, Bagram became a lifeline. In the beginning, living conditions were spartan, even bordering on primitive. Pilots and crews slept on cots, while meals were typically pre-packaged rations. Dust storms swept across the valley, coating everything with a fine layer of dust. But the hardships fostered a sense of shared purpose. Every sortie mattered. Every flight carried the weight of American and Allied troops fighting for their lives.

Over time, Bagram evolved from a patched-up ruin to the primary hub for NATO's air campaign. Runways were made serviceable, revetments were re-built, and maintenance tents gave way to semi-permanent hangars. Satellite uplinks connected the base to command centers across the globe. Yet even as the infrastructure modernized, the essence of 2001-02 lingered: Aircraft flying from a base that had been written off as rubble were now holding the line in a war without frontlines.

To this point, F-16 sorties were being launched primarily from the Gulf states—Kuwait, Qatar, and the UAE. This meant long flights via the Persian Gulf and Pakistan before the jets even reached Afghan airspace, with multiple refuels along the way. With ground forces pushing deeper into Afghanistan, and the Taliban retreating into their mountainous strongholds in the Hindu Kush, the US needed its fighters closer to the fight. Bagram, scarred and broken as it was, offered the best solution.

Still, it would be a few more years before the F-16 found a permanent hub at Bagram Air Base. Nevertheless, by establishing regular CAS operations from Bagram, the US

Air Force achieved more than just forward-basing. It had shortened the distance between decision and effect. By the time F-16s began flying from Bagram on a regular basis, they had seamlessly integrated into coalition's air power framework—delivering rapid-fire air support when and where it mattered the most. The daily operations at Bagram were, in many ways, the beginning of a long, grinding rhythm that would define air operations in Afghanistan for the next two decades.

Ironically, the first F-16 to touch down at Bagram wasn't American, and its arrival wasn't intentional. On December 19, 2002, a Danish F-16 diverted to Bagram after completing a strike mission. The Danish Viper was one of six deployed to Manas Air Base, Kyrgyzstan. As part of the growing NATO partnership in Afghanistan, the Royal Danish Air Force (RDAF) had deployed this small expeditionary force of half a dozen F-16AMs (Block 15 MLU variants) in October 2002.

The Bagram landing, however, was anything but pleasant. The pilot was on his final approach at just before 10:15 PM that night. Perhaps owing to fatigue, low visibility, or the inclement weather, the Viper skidded off the wet concrete of the northern runway, shuddering to a halt in a minefield 500 meters beyond the end of the airstrip. Major Stephen Clutter, a US Air Force spokesperson at Bagram, confirmed the RDAF Viper's crash in a statement to the Associated Press. "The jet, which had its front landing gear sheared off, overran the north runway," he said. "The pilot, who is from the Danish Air Force, was immediately evacuated to the US Army hospital here at Bagram and he was treated." The Danish pilot survived with only minor injuries, but his wrecked F-16 was another matter.

Aside from the structural damage (including a sheared targeting pod and a collapsed radome) he had plowed his bird into the middle of an active minefield. For any recovery crew, one misstep could be their last. The solution? The stricken Viper was hoisted from the deadly minefield by a CH-47 Chinook helicopter.

Although it wasn't the most flattering publicity for the RDAF, the incident nevertheless underscored how Enduring Freedom had morphed into a truly multinational effort.

By the summer of 2002, it was clear that Enduring Freedom had outgrown its single-nation origins. Invoking Article 5 of NATO's founding charter ("An attack on one, is an attack on all"), the battlefield grew exponentially as the air forces of Western Europe joined the fight. The F-16 was still at its center; but it was no longer exclusively an American asset. From the windswept tarmacs of Manas Air Base in Kyrgyzstan, a tri-national European detachment of F-16s—comprising Norway, Denmark, and the Netherlands—took shape. Their deployment marked the first true "expeditionary" air campaign in NATO's history.

The RDAF, alongside their Dutch and Norwegian counterparts, had deployed to Kyrgyzstan in October 2002. As mentioned above, the RDAF brought six F-16s carrying the MLU enhancements (one of which was lost during the ill-fated Bagram landing in December). The Norwegian and Dutch contingents likewise brought six Vipers each, rounding out the NATO delegation to 18 state-of-the-art Fighting Falcons, all backed by a Dutch KC-10 tanker. NATO F-16s operated in tight coordination but, unlike Operation Allied Force, these latter-day Vipers flew under distinct national chains of command.

Standard sortie packages included laser-guided GBU-12s, air-to-air missiles for any incidental self-defense, and targeting pods for nighttime CAS missions. Pilots were meticulously trained for pod-targeting operations and night-vision aided missions. Flights typically averaged six hours, with rotations every four weeks to manage pilot fatigue. Although the chains of command were rigid, a tight repair and maintenance network was forged across national lines. For example, a Dutch F-16 in Kyrgyzstan was recovered by a joint Norwegian-Danish rescue team.

By early 2003, these European F-16s had become more than just airborne symbols; they were multi-dimensional warfighters. Their integration helped solidify the shift from a US-led operation to a NATO-backed coalition.

In January of that year, the Royal Norwegian Air Force would see combat for the first time since World War II. Norwegian F-16s—even after two decades of service, thousands of flight hours logged, and a tour of duty in Operation Allied Force—had never fired a shot in anger.

That all changed on the morning of January 27, 2003. At the edge of the Adi Ghar Mountains in Kandahar Province, US Special Forces along with local Afghan troops stumbled upon a concealed force—nearly 80 Taliban and Al-Qaeda militants holed up in a series of cave complexes. A sudden firefight erupted, drawing helicopters and UAVs into the air. But as American forces pushed deeper into the fight, an urgent call came over the radio: "We need air."

This was Norway's moment.
The Norwegian F-16s, already orbiting their designated airspace, swooped in, dropping a pair of 2,000-lb GBU-12 bombs onto the Taliban positions. But for the troops on

the ground, it was only the beginning of a 15-day operation to clear the insurgents from their caves. The coalition called the effort to dig them out "Operation Mongoose," a deliberate, methodical fight built around Special Forces patrols and Afghan government troops, with devastating air power at the ready.

A few days after the Norwegian bombardment, Danish and Dutch F-16s followed suit, delivering their own precision bombs onto the cave complexes and other insurgent strongpoints. In total, coalition jets dropped nineteen GBU-2Ks and two GBU-12s, destroying more than 75 caves, and killing or capturing dozens of insurgents, all without a single coalition casualty. In terms of scope and intensity, Mongoose was the largest air-ground battle since Operation Anaconda.

All told, Mongoose was a tactical success for coalition forces. After-action reports highlighted destroyed caves, munitions, and logistical sites. Ground forces found weapons, fuel, and documents; and the enemy retreated farther south and east, stripped of their prepared sanctuaries inside the Adi Ghar mountain range. The pattern across those two weeks followed a familiar script: Ground forces moved, took fire, and called for air support. Then, F-16s and other fighters would answer the call with tight, controlled airstrikes.

It wasn't a headline-grabbing "decisive battle," but the cumulative effect of the operation mattered. Denying those sanctuaries reduced the enemy's capacity to stage and resupply. Moreover, it demonstrated how NATO partners could shoulder meaningful, lethal roles in Afghanistan's air campaign.

Inside Norway's expeditionary force, the moment carried tremendous weight. Back home, the news of their victory was well-received. The Royal Norwegian Air Force had flown CAPs over the Balkans in 1999, but those had been air-to-air missions with no demand for dropping bombs. But Afghanistan was different. It was a CAS-heavy war that trusted young pilots to put steel on target at razor-thin margins. Official Norwegian histories later described how their air force had hustled to field an autonomous air-to-ground capacity—precision weapons, procedures, and targeting pods—and how, during the early OEF rotations, a shortage of organic pods had forced careful integration with coalition laser teams.

As always, the pilots and crews relied on their training, discipline, and sheer grit. Together with the Dutch and Danish, the Royal Norwegian Air Force kept their small expeditionary footprint running at peak efficiency; and the entire European contingent adapted to an operational tempo where a single bad call could echo for months.

Following Operation Mongoose, the European F-16 detachment coalesced into the "European Participating Air Forces" (EPAF) team. By the spring of 2003, despite early friction from cultural and procedural differences, the three nations were increasingly working towards an integrated operational picture. In fact, commission-level reports praised joint action, even while divergent ROEs threatened to disrupt missions.

Over time, however, these complexities yielded to cohesion. By 2005, EPAF detachments were providing seamless CAS, surveillance, and "show of force" flyovers under the umbrella of NATO's expanding air campaign.

NATO's Counterinsurgency (2005-11)

From 2005 to 2009, the F-16 had become a fixture in the Afghan skies—Dutch, Norwegian, Danish, and American. Their bombs had cracked open Taliban strongholds in the Shah-i-Kot Valley, rained fire on insurgents within the Helmand Province, and blasted caves in Nuristan. By the winter of 2010, however, the war in Afghanistan had entered its bloodiest and most ambitious phase.

Nearly a decade of fighting had passed since the first US airstrikes in October 2001. Yet the insurgency had not been extinguished. Instead, it had evolved—splintering into networks of Taliban fighters, narco-traffickers, and foreign jihadists entrenched across the southern borderlands. For years, NATO aircraft had flown constant overwatch missions, supporting coalition patrols in Helmand and Kandahar.

Still, the Taliban endured.

By 2009, US commanders concluded that only a sweeping counterinsurgency (backed by President Obama's troop surge) could turn the tide. The mission parameters were broad and ambitious: (1) Clear and hold the Taliban-controlled regions; (2) Bring governance; and (3) Fracture the enemy's will to fight. At the center of this gambit was Operation Moshtarak (Dari for "Together"), a massive assault on the Taliban-held city of Marjah in Helmand Province. It would be the largest joint offensive of the war to date, involving 15,000 troops (Americans, British, Canadians, Estonians, and Afghan National Army soldiers). Overhead, the F-16 would once again prove indispensable.

In many ways, Operation Moshtarak was the product of lessons learned from 2005-09. The lessons from those four years had slowly reshaped NATO's aerial strategy. Indeed, by February 2010, the air war in Afghanistan was finely tuned. Aircrews had learned that overwhelming force could bring temporary victories, but lasting success required precision and restraint. The air war's ROE had tightened after incidents of civilian casualties in 2008, forcing pilots to coordinate more closely with ground forces and JTACs.

For Operation Moshtarak, the coalition orchestrated a layered air campaign. UAV drones scanned Marjah's irrigation canals and mud-brick compounds. C-130 and C-17 cargo planes airlifted supplies into forward bases. And in strike packages staged from Kandahar and Bagram Air Bases, F-16s stood by, ready to hit enemy bunkers, interdiction routes, or reinforcements moving along the Helmand River Valley.

Meanwhile, the Taliban, fully aware that the coalition was coming, spent weeks digging in. They seeded Marjah with mines and improvised explosive devices, built trench lines, and fortified compounds with interlocking fields of fire. To break through this layered defense, coalition commanders knew they would need airpower not just as support, but as the decisive edge.

In the pre-dawn hush of February 13, 2010, the assault on Marjah began. Heliborne Marines and Afghan soldiers descended onto the city's outskirts while British forces advanced from the north. Within hours, they met fierce resistance: Small-arms fire from compounds, RPG teams, and ambushes triggered by pressure-plate IEDs. The firefights were relentless, echoing across the city as dawn broke over Helmand.

Above them, F-16s from the 510th Fighter Squadron out of Aviano and Air National Guard units rotated into the fight. Their radios crackled with calls from JTACs embedded with ground troops from the 1st and 3rd Battalions of the 6th Marine Regiment. At one point during the assault, Taliban fighters had maneuvered into firing positions along irrigation ditches, pinning down a Marine squad in the northeast sector of Marjah.

An urgent appeal went up: "We need air on this target! Troops in contact!"

An F-16 roared in from above, banking hard to line up the strike. The pilot confirmed his coordinates, cleared the Marines' positions, and rolled in with his JDAM at the ready. The bomb zeroed in with pinpoint precision, obliterating a Taliban fire team while sparing nearby civilians. Over the course of that day, coalition F-16s dropped multiple precision-guided munitions, often under the tightest constraints. In some cases, instead of bombs, they executed show-of-force passes (with afterburners blazing) to scatter insurgents and break ambushes before they could develop.

For days, the F-16s orbited Marjah in relentless cycles. Pilots endured missions lasting 8-10 hours, necessitating multiple refuels from the KC-135 hovering over southern Afghanistan. The radio chatter was constant. Marines were calling for fire, British units were requesting ordnance on reinforced compounds, and Afghan soldiers asked for strikes on Taliban mortar teams.

One pilot later remarked that the intensity of Moshtarak rivaled the early years of Operation Enduring Freedom, but with more restrictive ROEs and a heavier reliance on precision strike coordination with JTACs.

By late February, coalition forces had secured Marjah's central districts, though the fighting would drag on for weeks. The Taliban dissolved into the countryside, their stronghold fractured but not destroyed. Still, Operation Moshtarak marked a turning point. It was the first large-scale offensive of the US troop surge, a demonstration that coalition forces could clear and hold enemy positions with Afghan partners in the lead.

For the F-16 community, Moshtarak reaffirmed the jet's role as the backbone of tactical air support in Afghanistan. From its earliest missions in 2001—when Vipers dropped bombs on Taliban camps near Kandahar—to the grinding battles of the troop surge, the F-16 had proven its adaptability. It could loiter for hours and respond within seconds to troops in contact, delivering the latest and greatest in precision-guided munitions. At Marjah, F-16s were the often-heard but rarely-seen guardian angels for Marines in the trenches. They were the veritable hammer against enemy strongpoints, and a guarantee to soldiers on the ground that no matter how intense the fight, air power was only a radio call away.

But Operation Moshtarak did *not* end the insurgency. Taliban fighters returned to contest Helmand highlands months later. But for the Marines who had taken Marjah, and for the pilots who risked long missions overhead, the mission stood as a paragon of joint resolve. F-16s had shown yet again why they remained the workhorse of NATO's air war in Afghanistan.

Reflecting on the success of the operation, General Stanley McChrystal, commander of the International Security Assistance Force (ISAF) in Afghanistan, noted in his 2010 commander's assessment: "The fusion of ground maneuver and precision airpower in Helmand exemplifies the coalition's ability to fight together—decisively and with discipline." And in the skies over Marjah, that fusion had ridden on the wings of the F-16.

Years of Transition (2012-15)

By 2012, the war in Afghanistan had taken on a different nature. Following Operation Neptune Spear in 2011, wherein the US Navy SEALs cornered and killed Osama bin Laden in Pakistan, NATO began its period of transition, gradually returning control of Afghanistan's security to the Afghan government. But even as NATO combat units drew down, the skies over Afghanistan did not empty. The Taliban insurgency was still afoot, and there was a lingering chance that they would re-seize control of Afghanistan once the coalition had departed. But for now, F-16s continued patrolling the high valleys and deserts, their JDAMs at the ready, in case ground forces needed help.

In fact, Air Force news outlets reporting from Bagram emphasized that, even in the war's waning years, the F-16s provided "24-hour overwatch and close air support" for NATO troops all over Afghanistan. Dutch Vipers also remained on station, flying CAS and conducting battlefield reconnaissance. In short, although boots on the ground were steadily reduced after 2012, NATO jets (American and Dutch) still flew armed patrols and stood ready, ensuring that Afghan and coalition soldiers never lacked the shield of air power.

In April 2014, Alabama's storied 100th Fighter Squadron—the modern-day "Red Tail" heirs to the World War II Tuskegee Airmen—touched down at Bagram Air Base. This Alabama National Guard unit assumed the mission of providing armed overwatch and close-air support for troops across Afghanistan. Normally, the Air

National Guard rotated its pilots and planes on shorter deployments, but the Red Tails' tour of duty was the first *six-month* deployment for a Guard squadron in Afghanistan. It reflected how, even as NATO forces withdrew, there was still a critical need for F-16 air power.

The 100th Fighter Squadron's Airmen understood the legacy they carried. In World War II, their squadron had flown bomber-escort missions over Italy. In 2009, F-16s from their parent unit, the 187th Fighter Wing, had flown combat sorties during Operation Iraqi Freedom. "It means a lot to be part of history," said Lieutenant Jeffrey Witt, proudly noting that the squadron had "flown with distinction during World War II" and was continuing that heritage in the mountains of Afghanistan.

Over the following months, the Red Tails launched daily missions across the country. On one mission, Lieutenant Colonel Mike McGinn and Major Ray Fowler were flying their typical on-call orbit, when a plea for help erupted over the radio. "We were out doing some base defense-type stuff," said McGinn, "watching over different sites, when we got the call...troops in contact about 40-50 miles away, which only takes us about five minutes to get there."

While en route to the beleaguered unit, the JTAC radioed in again: "Be advised, we've just had a medevac! Multiple Americans wounded!"

McGinn continued: "As soon as I keyed into the frequency to talk to the JTAC, you could hear the screaming in the background...hear the bullets firing...hear him breathing heavy because he's running." Then the JTAC uttered the words that no CAS pilot wants to hear:

"Danger Close!"

In that moment, Fowler and McGinn knew that they were going to be dropping bombs within just a few hundred meters of friendly troops. But McGinn realized they had another compounding problem. "The comms were terrible," he said, "I don't know if it's because his antenna was bouncing around as he's running down the streets." Whatever the case, McGinn and Fowler were each able to decipher about half of what the JTAC was saying.

Finally, during a brief moment of clear comms, the JTAC radioed: "Our people are getting shot up! We need those buildings dropped now!"—referring to the nearby village from which they were taking fire. But Ray Fowler needed to make sure he was dropping his bomb exactly where the beleaguered patrol needed it. Fowler radioed to McGinn: "He [JTAC] is clipping for me, and I'm not comfortable dropping this bomb until we find out exactly what they want in the middle of town." Without precise coordinates, he ran the risk of inflicting civilian casualties—or even worse, fratricide. "The worse thing for any fighter pilot," Fowler recalled, "or for anyone in this type of scenario, would be to hurt or kill friendly people on the ground."

But desperate times called for desperate measures. The JTAC radioed again: "Americans are going to die if you don't drop those f*cking bombs!"

That was all the confirmation Fowler and McGinn needed. They released their bombs and heard the muffled roar of the explosion. After a few dreadfully tense moments, the JTAC reappeared on the net, with an unmistakable calm in his voice:

"Viper, good effects."

Their bombs had neutralized the enemy force, with no collateral damage to friendly troops.

In practice, the pilots were constantly poised over areas of contact. One moment, they might be circling over the dusty roads of Kandahar, then racing north to scan the horizon over Kunduz the next moment. Throughout the six-month deployment, morale in the squadron ran high. As one intelligence officer noted: "We have met, if not exceeded expectations set forth by the previous squadron." Their performance was a true testament to the Red Tails legacy of superior airmanship under some of the most adverse conditions imaginable.

In many ways, the Red Tails' engagements were representative of the broader air mission over Northern Afghanistan in 2014. Several firefights were breaking out across the northern frontier, and coalition jets were increasingly being called upon to help. American F-16s at Bagram and Kandahar flew continuous patrol patterns over the north, ready to pick up any impromptu missions in support of coalition ground forces.

Dutch F-16s likewise orbited the Kunduz area to back up NATO trainers and Afghan soldiers. The Royal Netherlands Air Force confirmed that its jets were trained to give air support to friendly troops, and even use onboard sensors to detect roadside bombs. In fact, by late 2014, the Dutch F-16s—which had relocated north to Mazar-e-Sharif—had flown more than 10,000 sorties in Afghanistan before redeploying back to the Netherlands. Many of those flights were over the province of Kunduz. While exact engagements are seldom publicized, or overly-detailed in official histories, these verified reports leave little doubt: During the Kunduz battles, Allied F-16s were above the battlefield, providing watchful "eyes in the sky" above the troops in contact. Their presence—

whether dropping GBU bombs or simply roaring overhead in a show of force—was a critical lifeline for Afghan and NATO soldiers fighting in the fading year of 2014.

Final Years & Withdrawal (2015-21)

By late 2015, American F-16s were the only fighter jets remaining in Afghanistan. In October of that year, the 421st Fighter Squadron (of Desert Storm fame) arrived at Bagram for what would be their final F-16 deployment in country. Under the command of Lieutenant Colonel Michael Meyer, the squadron assumed the overwatch and CAS responsibilities for all remaining US and Afghan forces. As Meyer put it: "We are here to support the ground commander's intent...save lives on the ground and help transition Afghanistan to a stable and self-sufficient government."

Over the ensuing years, the F-16 rotations continued as Active Duty and Air National Guard squadrons took up the fight. For example, in July 2018, the South Dakota Air National Guard's 175th Fighter Squadron deployed four F-16Cs to Bagram. Remarks from their commander, Lieutenant Colonel Cory Kestel, reflected Meyer's remarks three years earlier. He emphasized the mission, while acknowledging the new, multi-layered enemy. In Kestel's words, the Squadron was to "support [coalition] forces and meet the ground commander's intent" to aid Afghan security forces "in their fight against Taliban, Al-Qaeda, ISIS, and other insurgents."

Likewise, the 510th Fighter Squadron returned for another round of CAS patrols in 2019. But by now, the mission had taken on a different flavor. Operation Enduring Freedom had officially ended in December 2014, and was replaced by Operation Freedom's Sentinel—a phased withdrawal wherein US troops would build up Afghan security forces so that the new government could "secure the Afghan people, win the peace, and contribute to stability throughout the region." Freedom's Sentinel itself was a nested mission underneath the larger NATO enterprise, coined Operation "Resolute Support," which had the same goals, albeit via training and advising the new Afghan forces.

Although the mission in Afghanistan was now in its twilight, many of the priorities remained unchanged. For example, the 510th Fighter Squadron's commander, Lieutenant Colonel Ben Freeborn, recalled that it was more important to "protect the population and support the government of Afghanistan than apply massive fire to the ground." Indeed, Freeborn praised his aviators for taking on extra risks to shield Afghan civilians.

Throughout 2016–2020, the American Vipers formed an enduring force. One commander noted that the F-16s "have had a constant presence in Afghanistan for more than a decade," with jets rotating through bases like Aviano, Italy, and Hill AFB, Utah. In practice, however, nearly all these squadrons were American assets. The coalition's last non-US F-16 contributions (from the European contingent) had ended years earlier. As the Afghan National Defense and Security Forces grew stronger, these fighter squadrons also helped "train, advise, and assist" the budding Afghan Air Force (AAF). American F-16Cs flew live-fires and joint exercises with AAF teams, even as the "Afghanization" effort aimed to hand over security roles to Kabul.

By early January 2016, the 421st Fighter Squadron was

already flying combat missions over Helmand Province. These F-16s struck Taliban-held positions around Lashkar Gah and the Nawa District, flying both day and night missions in support of US and Afghan forces. Staff Sergeant Robert Cloys, an Air Force public affairs specialist documenting one such sortie, noted that the squadron was the "only dedicated fighter squadron in the country, and continuously supports Operation Freedom's Sentinel and the NATO Resolute Support mission."

In the misty dawns and snowy flourishes of early 2016, crews from the 421st became the guardians of Helmand's "Ratchet Hills" and farmlands. Day after day, pilots peered through their helmet-mounted displays and targeting pods, ready to laser-designate targets. In Helmand's "green zone," where villages and orchards sat perilously close to Taliban outposts, the F-16 pilots improvised precision strikes. They dropped small bombs and called in Hellfire missiles from orbit, always mindful of the ground troops and villagers below.

By mid-January, with snow still blanketing parts of Helmand, the 421st had been a featured player in supporting the Marines from Camp Leatherneck, and several Afghan ground offensives. Indeed, the pilots maintained a loose ring of overwatch around coalition bases, relaying real-time targeting data to FACs on the ground. As the squadron's operational tempo peaked, they often flew the same routes night after night—their F-16s outfitted with night-vision-compatible cockpits and advanced targeting pods.

Nothing in Afghanistan's long campaign had required quite this mix of technology and grit. In Helmand's final winter under coalition control, these F-16s embodied the line between safety and chaos, knowing full well that every second they loitered above could mean saving lives below.

By late 2020, it was clear the war was winding down. The only question was how long the Afghan National Defense and Security Forces could keep a resurging Taliban at bay. Under the new peace deal and drawdown plans, American forces set a withdrawal deadline of September 2021. By the dawn of that year, American F-16s began closing out their final flights. By the spring of 2021, the 455th Air Expeditionary Wing quietly began packing up, sending its aircraft out of Afghanistan. On July 1, Bagram Air Base was abandoned without ceremony—a bitterly symbolic end to 20 years of US combat aviation in Afghanistan. Notably, no F-16s or other manned combat jets remained on the field.

Still, Afghan leaders and defense analysts warned that losing the Vipers' cover would be costly. In late July 2021, Taliban forces began sweeping across the northern and western provinces. Afghan President Ashraf Ghani personally appealed to the US for continued air support, saying that Afghan troops depended on it "to stem the Taliban advances." He reminded American officials that air power had been "key" to holding the line. Even the Pentagon later conceded that once the F-16s were gone, "the Taliban [were] out in the open," confident that their momentum could no longer be stopped now that American air power had departed. With US and NATO F-16s gone from the skies, Afghan pilots were left mostly with light attack aircraft and helicopters, a fraction of the air arm they had been promised.

In the end, the US rushed towards its withdrawal deadline amidst a storm of crises. On August 15, 2021 the Taliban thundered back into Kabul, completing the collapse that had quietly begun months before. By that time, all US F-16 operations had ceased—their pilots having flown some of the final combat sorties of a 20-year air campaign.

Thus concluded the era of F-16s in Afghanistan.

An F-16 pilot conducts an aerial refueling mission over the skies of Iraq during Operation Iraqi Freedom, November 2004. *US Air Force*

Chapter 6:
The Second Storm

Continuing the Global War on Terror, the United States soon turned its attention towards Iraq. Under the newly-articulated Bush Doctrine, the United States would pursue a policy of pre-emptive action against any country that harbored terrorists, or likely posed a threat to American security. "If we wait for threats to fully materialize," said President George W. Bush, "we will have waited too long."

But this march towards the "Second Gulf War" began long before the first coalition aircraft rocketed towards Baghdad in March 2003. Indeed, the seeds had been sown in the turbulent aftermath of Desert Storm twelve years earlier, when Saddam Hussein—bloodied but unbroken—remained defiant against international sanctions. Though his armies had been driven from Kuwait in 1991, Saddam remained in power, and continued his bloody reign of terror against the Iraqi people.

Meanwhile, throughout the 1990s, Iraq had become a land under siege. UN sanctions had cut deep into its economy. Yet Saddam maneuvered deftly, turning hardship into propaganda and maintaining his grip on the fractured Arab state. His defiance extended to the skies, where US and Allied fighters enforced the Northern and Southern No-Fly Zones, shielding the Kurd and Shiite minorities from reprisal while containing the despot's reach. A decade-long game of cat-and-mouse unfolded as Iraqi air defenses tested the Allied patrols.

And the F-16 was often the featured player in these games of aerial brinksmanship.

Tensions escalated further in 1993 when the US

launched cruise missiles into Iraq after Saddam was accused of plotting to assassinate former President George HW Bush. A few years later, the confrontation deepened when Iraq expelled UN weapons inspectors in 1997, prompting Operation Desert Fox in December 1998. The message was unmistakable: Saddam's defiance would not be ignored. Yet his regime endured, and its purported WMD program remained shrouded in secrecy and suspicion.

The turn of the millennium brought no relief. Saddam's cloak-and-dagger games with the UN continued, punctuated by sporadic clashes with Allied aircraft in the No-Fly Zones.

Then came the catastrophic events of September 11, 2001.

In the aftermath, the United States began looking for potential threats, no longer willing to gamble on deterrence alone. To Washington, Saddam's Iraq—with its history of aggression, use of chemical weapons, and unresolved questions about hidden stockpiles—represented a clear-and-present danger that could no longer be contained.

From late 2001 onward, the Bush Administration pressed its case. Iraq, they argued, was not just a regional menace but a possible nexus for terrorism and WMDs. President Bush declared Saddam's regime part of an "Axis of Evil," grouping it with North Korea and Iran as existential threats to world security. Diplomacy carried on throughout 2002, as UN inspectors returned to Iraq under intense scrutiny.

But their reports were inconclusive.

No smoking gun; but no clean bill of health either.

The United States and Britain insisted that Saddam's

evasiveness was sufficient proof of guilt, while other UN partners—particularly France, Germany, and Russia—argued for more evidence and greater patience.

By early 2003, however, the UN's patience had run out. In Kuwait, more than 250,000 American troops assembled alongside British and coalition partners. Naval strike groups descended onto the Persian Gulf, while air squadrons thundered into Saudi Arabia and the surrounding Gulf states. It was the largest Allied mobilization since Desert Storm, but with far greater ambitions: Not to repel an invasion, but to topple a regime, rebuild a nation, and restructure the balance of power in the Middle East.

The opening act of the air campaign would be dubbed "Shock & Awe." In the skies over Baghdad, precision-guided firepower rained down upon the regime's critical targets. And here again, the F-16 Fighting Falcon—battle-tested from a decade of No-Fly Zone patrols, SEAD operations in Yugoslavia, and its more-recent CAS enterprises in Afghanistan—stood at the forefront.

In the final year before the invasion, the Gulf's airfields felt like the staging grounds of an oncoming storm—sunbaked airstrips lined with F-16s that had been staring down Iraq for a decade, and now found themselves flying in a two-front war. Prince Sultan Air Base in Saudi Arabia had become the central hub, where the 363rd Air Expeditionary Wing ran the daily grind of Operation Southern Watch while doubling as the Combined Air Operations Center (CAOC) for the new war in Afghanistan.

Around the Gulf, F-16 units rotated like clockwork under the expeditionary model. At Kuwait's Ahmad al-Jaber Air Base, the 332nd Air Expeditionary Group hosted a rolling cast of Active Duty and Guard/Reserve Vipers for Southern Watch, then pushed elements into Enduring Freedom as the Afghan campaign got underway.

Even the jets themselves had undergone some radical changes. The Common Configuration Implementation Program (CCIP) began filtering into frontline squadrons during the 2001–2002 season—the largest American F-16 upgrade to date. The CCIP unified *Block 40/50* cockpits and avionics, enabled Link 16 data-link connectivity, Joint Helmet-Mounted Cueing, and broader weapons integration—capabilities that would pay dividends when the OSW routine gave way to the opening rounds of Iraqi Freedom. Meanwhile, the Air National Guard and Reserve F-16s had fielded the LITENING pod in 1999, with enhanced variants delivered to the expeditionary units throughout 2002-2003. Paired with JDAMs, JSOWs, and WCMDs, these Air Guard fighters brought precision-on-demand for the daily No-Fly Zone patrols and the asymmetric fights over Afghanistan.

For the Wild Weasel F-16s (Block 50/52 CJ variants from the Balkans-era), the HARMs Targeting System (HTS) kept evolving. The Revision-7 (later STING) path added a digital receiver and the ability to pass geolocations over Link 16, converting radar blips into coordinates that any GPS-guided weapon could use. In a region still teeming with Iraqi SAMs and air defense guns, these upgrades meant that the Wild Weasel was no longer just a shepherd for strike packages; it could cue precision fires across any formation and prosecute targets by itself.

Operationally, 2002 felt like a hinge. OSW still demanded its daily ritual: "Sand runs" up to the 33rd Parallel, reactive strikes on radar emitters, and a constant presence to suppress Iraqi air defenses. But the same squadrons were now flying missions for Enduring Freedom, too—some from Kuwait and Qatar, but all under the CAOC's widening scope—stretching their arms across *two* major theaters with two separate mission parameters.

However, the first moves of the air campaign against Iraq began not on the flight lines, but in the Air Operations Centers where strategies were made. At Ramstein Air Base in Germany, Major Anthony Roberson, a USAF Weapons Instructor and seasoned F-16 pilot, found himself at the heart of these preparations with the 32nd Air Operations Group. By mid-2002, as Washington's rhetoric hardened and the war plans accelerated, Roberson's task was to inject a fighter pilot's perspective into the Joint Air Operations Plan. His responsibility was monumental: Help draft the script for an air war that would topple Saddam Hussein.

"We had the first 30 days planned 90 days before the war started," Roberson recalled, highlighting the meticulous level of detail needed for operational planning—"what bomb, on what target, at what time."

From the planning cells at Ramstein, he worked to ensure the air campaign would not be southern-centric, balancing the weight of forces in Kuwait and Saudi Arabia with efforts to open a northern aerial route through Turkey. The plan accounted for everything: Insertion of Special Operations forces, seizing and securing vital oil platforms, and the timing of conventional ground force movements into Iraq. For Roberson and his team, the challenge was not just to plan the air war, but to orchestrate the opening rounds of a bold, ambitious regime change.

By late 2002, the familiar rhythms of the No-Fly Zone had given way to a steady escalation of deployments and combat rehearsals. From these developments, one thing was clear: The Viper community was preparing for war.

The Opening Gambit (Spring 2003)

On March 19, 2003, the world watched as the skies above Iraq erupted with the opening salvos of Operation Iraqi Freedom (OIF). Beneath the thunder of Tomahawk missiles and the roar of heavy bombers, American F-16s fought a quieter but equally-decisive war. They flew into battle carrying precision weapons, cutting-edge avionics, and the burden of a mission that demanded nothing less than absolute dominance.

Theirs was not a single battle, or even a single front. Each unit bore its own weight, its own hardships, and its own victories. Together, however, they composed a mosaic of modern air power—a Viper force whose relentless fury shattered Saddam's defenses, crippled his forces, and cleared a path for the race to Baghdad.

The initial air campaign was designed to be overwhelming. It was the much-heralded "Shock & Awe"—a fury of precision-guided bombs intended to paralyze Baghdad's command structure and shatter enemy morale. For the F-16s, the mission was both new *and* familiar. Familiar because they had been the workhorses of Southern Watch for more than a decade; new because never before had these Vipers wielded such a high-tech arsenal in combat.

This is their story.

77th Fighter Squadron

Stationed under the 363rd Air Expeditionary Wing at Prince Sultan Air Base, the 77th Fighter Squadron was a longtime veteran of Southern Watch. But by early 2003, their mission was changing. That shift came gradually, yet with unmistakable clarity. As one pilot recalled: "It was pretty clear from the start that this wasn't just another Southern Watch rotation. The whole tempo was different." Every long sortie over the Southern No-Fly Zone now doubled as reconnaissance, testing routes and building the situational awareness that would soon matter in combat.

Their F-16CJs, equipped with AGM-88 HARMs and precision-guided bombs, were tasked with suppressing Iraq's air defenses and striking key regime targets in Baghdad. The first nights of war would test their training and resolve.

But the squadron did not stand alone. Alongside it were other units pulled from the farthest reaches of the Air Force's Viper fleet. This included the 14th Fighter Squadron from Misawa Air Base, Japan, with their own F-16CJs; a six-ship detachment of *Block 40* Vipers from the 4th Fighter Squadron, and an additional six *Block 30* F-16s from the Air Force Reserve's 457th Fighter Squadron. Together, they formed the backbone of the Expeditionary Wing's ground-attack capability. Yet it was the 77th that stood at the forefront, charged with defending the battle space and safeguarding the coalition strike packages.

During the final days of Southern Watch, the 77th flew missions that mirrored the long vigilance of the preceding decade. They escorted high-value reconnaissance platforms such as the U-2, prowling the borders of Baghdad's Missile Exclusion Zone (MEZ). Their F-16CJs, fitted with the HTS, listened for the telltale signals of Iraqi radars. At times, the air defense batteries tested them, firing unguided or optically-guided weapons into the skies. Yet for the most part, Saddam Hussein's defenses remained curiously restrained.

When OIF began on March 19, the 77th was already poised for the opening rounds. The squadron's pilots took their place in escort formations, each loaded with AGM-88 HARMs, flying close guard over the strike packages inbound for Baghdad and beyond. Their mission was simple: Keep the skies clear of threats, strike down hostile radar emissions, and ensure that no Allied bomber or strike fighter fell prey to Iraqi defenses.

The MEZ around Baghdad was a fortress of overlapping threats, with hundreds of SAM batteries spread throughout the city and its suburbs. Allied Intelligence couldn't precisely map them all, and many remained elusive. It was into this contested airspace that the 77th Fighter Squadron flew, often paired with coalition strike packages that included RAF Tornadoes, USAF bombers, and Navy/Marine strike aircraft. On March 22, for example, a four-ship flight from the 77th fired multiple AGM-88s against SA-2 sites at the Al Taqaddum Air Base, northwest of Baghdad, in support of the vanguard RAF Tornados. It was the opening act of a violent campaign against the enemy's air defenses.

For the young pilots of the 77th, OIF was a baptism by fire. Some arrived in theater with fewer than 200 flight hours aboard the Viper, thrust into missions that demanded both skill and composure.

Yet the squadron held together.

Day by day, they did the unglamorous work to grind down Iraq's modern air defenses. They hunted SAMs in the suburbs, covered the aerial seams between strike routes, and took pop-up taskings to support Allied aircraft punching into the city. Their operations blended HARM shots with time-sensitive moves, always under the knowledge that the MEZ's map was incomplete and shifting. When required, they surged into CAS and interdiction, including late-war "shows of force" to stabilize liberated areas, and targeted strikes on incidental targets along the approaches to Baghdad and in Tikrit.

By the fourth day of the air war, it became clear, however, that Saddam's air defenses were not behaving as anticipated. Rather than activating their radars and attempting to contest the skies, many of Saddam's SAMs opted to stay silent, relying instead on the occasional unguided launch in hopes of landing a lucky punch. It was nearly identical to the tactics they had employed during the waning days of Southern Watch. Still, the tactical games of "playing possum" posed a paradox for the SEAD mission. Suppression was achieved by mere presence. In other words, the threat of the F-16CJ and its HARMs were often enough to keep enemy radars offline. But the coalition's objectives demanded more. Strike aircraft had to penetrate the airspace into Baghdad and Tikrit to deliver their bombs. Thus, if the SAMs remained hidden in silence, they could easily spring to life against the strike aircraft once the Wild Weasels had vacated the airspace.

The solution, therefore, was a shift in doctrine.
The 77th, with their reputation for aggressiveness and skill, were chosen to spearhead the transition from pure suppression (SEAD) to active destruction (DEAD). Their aircraft began to carry mixed-ordnance loads: Some jets armed with HARMs; others with JDAMs, Wind-Corrected Munition Dispensers (WCMDs), or AGM-65 Mavericks. Now, the F-16CJs would not merely *suppress*; they would hunt and *destroy*.

The missions that followed unfolded in an atmosphere of uncertainty and danger. Strike packages often came under fire from air defense guns and optically-guided missiles as they pressed towards their targets. Yet these F-16s flew with determination, striking radar sites, missile batteries, and defensive positions. On March 26, one mission saw a pair of F-16CJs drop their WCMDs through cloud cover onto a SAM site south of Baghdad, their weapons guided by inertial navigation to erase the hidden threat. Other missions involved racing at full afterburner to support bombers striking Saddam International Airport, or to silence any variety of defenses around the capital.

By late March, however, the air campaign's focus had shifted from Shock & Awe to direct support of ground troops. As US Army and Marine Corps units surged towards Baghdad, the Vipers became their constant guardians. Equipped with LITENING and LANTIRN pods, F-16s could relay imagery, designate artillery targets, or deliver their own firepower in support of embattled units. Such missions demonstrated the squadron's adaptability, able to shift from SEAD to CAS at a moment's notice.

But as April dawned, the campaign entered its decisive phase. By this time, the Wild Weasels had reduced Iraq's air defenses to a hollow shell. As US ground forces pressed north towards Baghdad, Saddam Hussein's heralded Medina Division of the Republican Guard attempted to reposition. A massive armored column of nearly 1,000 vehicles had been spotted south of Karbala, threatening to stall the American advance. Captain Gene Sherer, an F-16 pilot aloft that day, recalled: "As soon as I hit the clouds at 15,000 feet, I was about seven-degrees nose down, and it was about as uncomfortable as I've ever been in an F-16. It was about [4:00 PM], so it was relatively bright, but I flew into a patch of cloud that firstly got dark and then turned dark red. At that point, I got lit up by a threat radar and recovered about six nautical miles from the bad guys."

But even amid the chaos of bad weather, radar threats, and poor visibility, the F-16 packages struck with devastating precision. Mixed-load sorties pummeled the armored column, halting its attempt to block the US Army's drive northward.

The result was decisive.

The Medina Division's counterattack collapsed under relentless Viper strikes.

By the time Baghdad fell on April 9, 2003, the 77th Fighter Squadron had flown hundreds of combat sorties into Iraq's most heavily defended skies. Statistics of the OIF air campaign tell the story of a squadron that carried a heavy burden during the opening rounds, and then adapted as the ground phase closed in towards the capital. Over the course of 676 sorties and more than 3,800 combat hours, the squadron's F-16s expended 170 CBU-103s and 105 AGM-88 HARMs, plus 52 JDAMs, 16 AGM-65 Mavericks, and more than 7,000 rounds of 20mm ammunition. In total, the squadron engaged 338 ground targets—including 104 air defense pieces (SAMs, radars, and anti-aircraft guns) destroyed or disabled, along with 20 armored vehicles, 26 trucks, and 36 aircraft on the ground. The squadron had absorbed the lessons of SEAD, adapted to the necessity of destruction, and emerged bloodied but unbroken.

Thus, the squadron's role in OIF was both tactical and symbolic. They represented the sharp end of the coalition's drive to dominate the skies; and their F-16s thrust deep into hostile territory from the very first night. Within the broader scope of the 363rd Air Expeditionary Wing, the 77th stood as the vanguard of fire—a guardian of the Allied air forces, and one of the decisive instruments by which the coalition won the skies over Iraq.

379th Air Expeditionary Wing

Across the sands of Qatar, at Al Udeid Air Base, rose a bastion of air power that would become one of the central nerve centers of Operation Iraqi Freedom: The 379th Air Expeditionary Wing. Its sprawling flightline would soon be filled to capacity as the Vipers from Spangdahlem Air Base and their allies arrived in the opening weeks of 2003. Here, the F-16CJs Wild Weasels of the 379th took their place as multirole predators in the skies over Iraq.

Ironically, the genesis of this deployment stemmed from aborted diplomatic plans. The F-16 squadrons of the 52nd Fighter Wing (the 22nd and 23rd Fighter Squadrons) had originally been intended for Incirlik Air Base in Turkey, to open a northern front in accordance with the coalition's design for a multi-axis campaign. But when negotiations with the Turkish government failed, the Spangdahlem squadrons were ordered south to Qatar. On January 17, 2003, twelve aircraft from the 22nd Fighter Squadron lifted off from Germany and touched down at Al Udeid. The 23rd Fighter Squadron joined them the following month, bringing another dozen CJ variants into theater, and thereby swelling the Viper presence to full strength.

Still, the relocation brought challenges. Diplomatic clearance for combat operations had to be painstakingly negotiated with Qatari authorities.

In the meantime, crews and aircraft crowded the base, swelling the Allied presence at Al Udeid into a hive of activity. Alongside the Spangdahlem squadrons came the 157th Fighter Squadron of the South Carolina Air National Guard, equipped with identical *Block 50* CJs. Together, these three units formed a composite force under the banner of the 379th.

Their mission set was broad and uncompromising: The 22nd, 23rd, and 157th were tasked with suppressing/destroying enemy air defenses, time-sensitive targeting, and CAS. From the first nights of OIF, their F-16CJs flew into Iraq in tight coordination with strike packages bound for Baghdad, Tikrit, and beyond. Each aircraft carried HTS-guided HARMs, ready to fire at the first flicker of hostile radar emissions.

For the 379th's composite units, the final days of Southern Watch had been a proving ground. The 23rd Fighter Squadron logs, for example, recorded 129 sorties flown during those weeks, tallying more than 600 hours of combat time before OIF even began. These missions gathered intelligence on Iraqi defenses, tested the discipline of Saddam's radar crews, and gave the coalition a clearer picture of the threats that would have to be dismantled once the air war commenced. The Spangdahlem squadrons also flew force protection missions, standing guard over bombers and strikers that launched punitive strikes against radar sites, observation posts, and command centers along the Jordanian frontier.

By March, the transition to full-scale war was complete. The 379th's Vipers were among the first into hostile airspace on March 19, just as their Southern Watch mission *officially* converted to Iraqi Freedom. That very afternoon, F-15Es dropped the last bomb of the No-Fly Zone era, and the curtain lifted on the wider campaign of OIF.

The role of the 379th's composite F-16 fleet was central to the Allied strategy. Whereas the 77th Fighter Squadron at Prince Sultan focused on suppressing Baghdad's defenses, the Spangdahlem squadrons were thrust directly at the heart of the Super MEZ. Their aircraft penetrated the very core of Iraq's air defense architecture, targeting radar nodes and SAM sites that ringed the capital. Every sortie demanded precision, coordination, and endurance. They fired HARMs in the opening minutes of every mission, sometimes against active radars, sometimes in pre-planned mode to suppress sites that might awaken as the strike packages approached.

Yet, as the days passed, the same paradox faced by the 77th's Wild Weasels revealed itself. Iraqi radar crews were staying mum. The silence of the air defense network—whether born from fear, command directives, or confusion—denied the CJs their opportunity to strike in traditional SEAD fashion. Coalition commanders, recognizing that suppression alone could not guarantee freedom of maneuver, authorized the shift towards destruction. The Spangdahlem and South Carolina CJs thus began carrying mixed loads, combining HARMs with JDAMs and cluster munitions. When a SAM site was identified, the directive was no longer to silence it but to obliterate it outright.

Night after night, the squadrons pressed into the densest concentrations of SAM and anti-aircraft defenses. Their pilots flew among the tracer rounds erupting from Baghdad's air defense guns, threading through skies ablaze with unguided launches. Against this chaos, they employed their weapons methodically, erasing SAM batteries, knocking out radar posts, and clearing corridors through which the bombers and strike fighters could pass.

Flexibility remained a hallmark of their operations. Beyond suppression and destruction, the 379th's F-16s also performed time-sensitive targeting, striking when intel identified high-value targets on short notice. They answered calls for CAS, delivering the same precision-guided fires that were protecting coalition ground forces elsewhere along the battlefront. As expected, their targeting pods made them invaluable in locating and destroying mobile assets or defensive positions that threatened the advance.

The tempo was relentless. The ramps at Al Udeid were packed shoulder-to-shoulder with combat aircraft, fully-armed Vipers nestled among tankers, bombers, and surveillance platforms. Maintenance crews worked without respite, turning aircraft in hours so they could launch again into the night. For the pilots, each mission brought the possibility of chaos—whether a radar suddenly springing to life, a missile arcing through the sky, or a call from troops in desperate need of fire support.

Through it all, the 379th's F-16s remained unwavering. Their presence in the skies above Iraq was both deterrent and destruction. By the campaign's end, the squadrons of Spangdahlem and South Carolina had flown hundreds of combat sorties, fired HARMs into the heart of the Iraqi defense network, and delivered precision strikes that smothered any hope of coherent resistance.

In the annals of the air campaign, the 379th Air Expeditionary Wing stood as a citadel in the desert. From its crowded runways surged a torrent of combat power, its F-16CJs the very embodiment of suppression and destruction. Their missions secured the skies for the coalition's advance, dismantled the defenses of Baghdad, and carried the war to the very heart of Saddam's regime.

524th Fighter Squadron

From the runways of Al Jaber Air Base in Kuwait, the 524th Fighter Squadron prepared to carry the fight into Iraq. Drawn from Cannon AFB in New Mexico, they arrived in theater with their *Block 40* F-16CGs in late 2002, initially as part of the Southern Watch rotations, then augmented for the war that loomed on the horizon. Eighteen Vipers, their fuselages bearing the marks of long years of vigilance, stood ready in the desert, forming the striking core of the 332nd Air Expeditionary Wing.

The 332nd AEW carried a name steeped in heritage. Its lineage traced back to the Tuskegee Airmen of the Second World War, whose red-tailed fighters had once carved a legend in the skies of Europe. Now, in the deserts of the Middle East, that banner flew again, a symbol of continuity across generations. To the pilots and maintainers of the 524th, their place in that lineage added weight to every sortie they flew.

Their mission was as diverse as it was demanding. Positioned in Kuwait, they were tasked first with guarding the border itself, patrolling the invisible line where Saddam Hussein's forces loomed just across the sand. They scoured the frontier for incursions, providing what commanders termed "Non-traditional Intelligence, Surveillance, and Reconnaissance" (NTISR). With targeting pods slung beneath their intakes, the F-16s scanned for enemy movements, their sensors capturing signs of Iraqi armor, artillery, and defensive positions. The intelligence thus gathered gave coalition leaders a vital edge in the final weeks before the war.

The squadron's aircraft also carried out a quieter mission in those last days of Southern Watch: Psychological operations. From their pylons fell leaflets, scattering messages across Iraq's southern units, urging surrender and sowing uncertainty. These drops, while less dramatic than missile strikes or bomb runs, played their part in shaping the battlefield before a single shot of OIF was fired.

When the war began, the 524th was thrust directly into combat. Their Block 40 F-16CGs, equipped with LANTIRN pods, were among the most capable precision night-strike aircraft in the theater. Tasked with close air support, interdiction, and deliberate strikes, they bore much of the burden in clearing the path for Allied ground forces as they surged north from Kuwait. Their missions ranged from border sweeps to deep strikes against airfields and command posts, to attacks on armored units that sought to block the coalition advance.

One mission, flown on April 2, encapsulated the squadron's role. Two aircraft, circling at altitude over Baghdad International Airport, methodically dismantled Iraqi defenses. In racetrack orbits, they dropped precision-guided bombs on bunkers, tanks, and fortified positions. Their strikes helped soften resistance at the gates of Baghdad itself, clearing the way for coalition ground forces to seize the airfield within hours. For their actions that day, the pilots were awarded the Distinguished Flying Cross, an emblem of valor in the skies.

But such highlights were only fragments of the 524th's relentless effort. Day after day, sorties flowed from Al Jaber, the aircraft returning caked in sand and soot, only to be rearmed and launched again. The tempo was unyielding. Maintenance crews toiled through the night to keep jets serviceable, their hands blistered and uniforms streaked with oil. Pilots briefed, flew, debriefed, and briefed again, their bodies pushed to the limits of endurance by the pace of combat.

The diversity of the squadron's missions reflected the expeditionary ethos of the 332nd AEW itself. The Wing was built from disparate elements: Active-duty squadrons, Air National Guard units, and reservists—all fused into a single force. The 524th's presence in the wing underscored its importance: a squadron equally capable of precision strike, border defense, and direct support to troops under fire.

The battles they fought were not always against fixed targets. As the war unfolded, the 524th's F-16s were increasingly tasked with supporting ground units under sudden attack. Calls came over the radios from armored columns pushing north, sometimes pinned by Fedayeen fighters or by desperate Iraqi counterattacks. When those calls reached the 524th, the response was swift. Within minutes, Vipers rolled in, delivering cluster munitions or laser-guided bombs with devastating effect. For soldiers on the ground, the roar of the Fighting Falcon overhead often marked the turning of the tide.

By the time Baghdad fell, the 524th had flown an extraordinary volume of combat sorties. They had patrolled the Kuwaiti border, hunted tanks and artillery in southern Iraq, suppressed enemy defenses around the capital, and lent their firepower directly to embattled units. They had delivered leaflets *and* bombs, shifting seamlessly between the psychological and the kinetic instruments of war.

Their record of achievement was not measured in a single dramatic engagement, but in the cumulative weight of their efforts—the steady, relentless hammering of targets across the campaign. The squadron embodied the principle of persistence. In every phase of the war, from the prelude of OSW through the lightning dash to Baghdad, the 524th's aircraft were present.

The red-tailed legacy of the 332nd AEW thus found new expression in the desert war of 2003. The Tuskegee Airmen had once escorted bombers over Europe; now their descendants cleared the skies and battlefields over Iraq. The 524th Fighter Squadron, as the wing's cutting edge, proved worthy of that inheritance. In their precision, their endurance, and their unyielding presence, they struck as a spear against the defenses of Saddam Hussein's regime.

410th Air Expeditionary Wing

Beyond the main thrust of coalition air power stood an armada shrouded in secrecy: The 410th Air Expeditionary Wing. Unlike the sprawling bases in Saudi Arabia, Qatar, and Kuwait, where fighter jets filled every ramp to capacity, the 410th Air Expeditionary Wing took shape on an austere fighter base in Azraq, Jordan. Its presence was the product of delicate diplomacy; Jordan's leaders had reluctantly agreed to host American combat aircraft. Yet once established, the 410th assumed one of the most politically-sensitive and strategically vital roles of the campaign: To watch, to strike, and to deny Iraqi aircraft any sanctuary in its vast western desert.

The 410th was unique among the coalition's expeditionary wings. Its backbone came not from the Active Duty force, but from the Air National Guard and Air Force Reserve. At its core was the 160th Fighter Squadron (Alabama Air National Guard), whose *Block 30* F-16s carried the enhanced AAQ-28 LITENING II targeting pod, among other modern systems. This pod, rare among Air Guard units, transformed the jet into a potent reconnaissance and strike platform, earning it the unofficial designation "F-16C+." Alongside Alabama flew the 120th Fighter Squadron of the Colorado Air National Guard, the 121st Fighter Squadron of the DC Air National Guard, and the Air Force Reserve's 466th Fighter Squadron. Together, these disparate squadrons formed a composite wing whose reach would stretch across Iraq's western frontier.

Their mission: To prevent Saddam Hussein from launching his Scud missiles, and to deny his regime any avenue of escape through the deserts leading into Syria. The threat of Scud launchers had haunted coalition planners since the Gulf War in 1991. Given the politically-tenuous nature of the 2003 coalition, the Allies could not allow a repeat, much less allow Saddam to goad Israel into the conflict. Thus, the 410th took up the role of anti-theater ballistic missile (anti-TBM) patrols, a ceaseless watch over the barren expanses of western Iraq.

In the weeks before the war, the 410th's sorties helped lay the groundwork for OIF. Their intelligence-gathering flights identified Iraqi defenses, thus shaping the battlefield for the Special Forces teams that would move in once the campaign began. When the war opened in March, the Wing shifted seamlessly from watchfulness to combat, patrolling the skies, responding to time-sensitive targets, and ensuring that no missile threats emerged to destabilize the coalition's advance.

The geography of their mission was stark. The desert stretched wide and empty, cut by a few roads, dotted with observation posts and radar sites. Within this vastness, mobile Scud launchers could lie in wait for any opportunity to strike. To counter them, the 410th's F-16s flew long sorties at low and medium altitude. With their LITENING II pods scanning for movement, the Air Guard pilots watched with deliberate intent, waiting to detect the faintest signatures of a camouflaged launcher. It was a mission that required as much patience as it did power; a mission where maintaining vigilance would be as decisive as ordnance.

But the 410th's efforts did not stand alone. On the ground, Special Forces teams continued their deep infiltration into the desert, hunting Scuds and relaying intelligence. In the skies above, the F-16s formed their "anti-TBM umbrella," providing a shield and sword for operators on the ground. Their presence in the skies meant that any sightings of an enemy launcher would be met with precision-guided firepower. Their targeting pods, meanwhile, fed a steady stream of recon imagery back to the 410th's command center, mapping radar sites, observation posts, and potential escape routes.

The conditions of their deployment were spartan. Indeed, Azraq lacked the infrastructure of Al Udeid or Prince Sultan. Within this austere environment, their F-16s operated from a base crowded with RAF Harriers and Canberra PR9 reconnaissance aircraft. Yet from this remote outpost, their influence stretched deep into Iraq. Every sortie flown over the western desert extended the coalition's control, sealing off an entire section of the country from enemy activity.

The importance of their mission lay not only in preventing Scud launches, but in shaping the strategic environment of the war. By denying Saddam the ability to threaten Israel or Jordan with incoming Scuds, the 410th closed off avenues of escalation that might have fractured the coalition along geopolitical lines. By scouring the desert for signs of movement, they limited the dictator's ability to flee toward Syria. And by supporting Special Forces patrols, they enabled the ground effort that dismantled the Scud units before they could act.

Theirs was not a mission of heart-pounding dogfights or dramatic SAM dodges. Rather, it was a campaign of watchfulness, endurance, and precision. Yet the 410th's impact was profound. By the time Baghdad fell, the western desert remained silent. No Scuds in the sky; no ballistic threats to unsettle the coalition's advance. The 410th Air Expeditionary Wing, in its secrecy and isolation, had unequivocally succeeded in its charge.

The legacy of the 410th in OIF was one of quiet decisiveness. From its hidden base in Jordan, it ensured that the war remained contained, the skies remained clear, and that Saddam's last gambits were denied. These Guardsmen and Reservists had proven the value of citizen-airmen in modern war—skilled, adaptable, and unyielding.

In the greater tapestry of the air campaign, the 410th Air Expeditionary Wing was a sentinel on the flank, guarding the coalition's vulnerable western airspace. Its F-16s patrolled the skies with tireless resolve, their presence as vital as the bombs they carried. And in so doing, they closed the doorway through which Saddam Hussein's regime might have struck back or escaped.

Baghdad fell to coalition forces on April 9, 2003. Television cameras captured the iconic image of Saddam's statue toppling from its pedestal in Firdos Square. But in the skies above, American F-16s continued to orbit, their presence a reminder of how air power had paved the way to victory.

By the end of major combat operations, F-16 squadrons

had flown thousands of sorties, validating the Viper's transformation into a multi-faceted, precision-strike platform. They had suppressed enemy air defenses, shattered command centers, shielded ground forces, and silently hunted Scud launchers on the western frontier. Together, these amalgamated Viper units embodied the true spirit of expeditionary warfare: Diverse in origin, united in mission, and relentless in execution. From the opening salvos to the fall of Baghdad, these F-16s filled the skies with precision and resolve.

But even as the headlines proclaimed "Mission Accomplished," the Viper pilots knew their war was far from over. Ahead of them lay eight years of insurgency and counterinsurgency. But the opening campaign had already carved the F-16's place into the history of Operation Iraqi Freedom.

Asymmetric Storm (2003-06)

With the fall of Baghdad in April 2003, the thunder of the initial invasion quickly faded into silence. The lightning advance of coalition forces, supported by laser-focused air power, had broken the organized resistance of Saddam Hussein's regime.

Yet the silence that followed wasn't "peace."
Across Iraq, a violent insurgency took root in the rubble of the dictatorship, fed by chaos, vengeance, and transnational jihadists. But even in the chaotic aftermath of Saddam's collapse, the skies remained as vital as ever. American F-16s, once the scourge of Iraqi air defenses, now settled into a long vigil that would drag on for years.

The post-conventional phase demanded adaptation. Missions that had once focused on SEAD and high-altitude bombing gave way to a new pattern of operations: CAS for coalition ground forces, deterrence patrols, rapid-fire responses to fleeting targets, and the emerging discipline of NTISR. To the pilots and crews, the message was clear: The air campaign hadn't ended; it had simply changed its character.

From Kuwait, Qatar, Saudi Arabia, and Jordan, the expeditionary wings rebalanced their presence. Many of the squadrons that had spearheaded the invasion remained in theater, their tours extended into the uncertain summer months. Others rotated out, replaced by incoming units drawn from the stateside wings and the Air National Guard. What had begun as a surge of combat power became an enduring rotation, an air bridge across the Atlantic to sustain operations in the skies over Iraq.

The 77th Expeditionary Fighter Squadron, champions of the SEAD campaign over Baghdad, continued to fly, but their missions had now turned to supporting convoys, patrolling restive cities, and providing overwatch for ground operations. Their *Block 50* F-16CJs, once bristling with AGM-88s and the vaunted HTS, now more often carried weapons tailored for narrow-margin CAS runs.

Back at Al Udeid, the 379th Air Expeditionary Wing likewise shifted its emphasis. The Spangdahlem squadrons and their South Carolina partners transitioned from dismantling SAM belts to providing consistent surveillance over Iraq's heartland. Their targeting pods, once used to cue HARMs, had now been repurposed for reconnaissance. Under the new parameters, they orbited overhead as the "eyes in the sky" for local ground commanders. The tempo of combat shifted from the concentrated fury of March-April to the grinding demands of daily patrols over several months, where success was measured by the absence of insurgent attacks.

At Al Jaber, the 524th Fighter Squadron and the 332nd Air Expeditionary Wing adapted just as quickly. Their jets, honed for precision night strikes during the invasion, were now tasked with hunting insurgent cells, protecting coalition outposts, and eliminating weapons caches. They flew over Baghdad, Fallujah, and Mosul, often coordinating directly with ground forces in contact. At times, they were called upon to strike buildings used as safehouses, insurgent vehicles, or fortified positions that were firing on coalition troops. Their pilots mastered the art of loitering overhead, ready to strike at a moment's notice.

The 410th Air Expeditionary Wing, still stationed at their austere base in Jordan, maintained their focus on Iraq's western desert. But now that Saddam's regime had fallen, the mission evolved to interdicting smugglers and preventing insurgents from using the desert as a sanctuary. Their vigilance ensured that Iraq's western frontier remained safe from infiltration, while their long-duration sorties provided reconnaissance that would shape operations far beyond their desert patrol lines.

Throughout the summer of 2003, as major combat operations were declared over, the coalition's air posture settled into an enduring pattern. Expeditionary fighter squadrons rotated in and out of theater on a regular cycle. The Air National Guard and Air Force Reserve played a vital role in sustaining that tempo. Indeed, units from across the continental US—many of which hadn't seen combat since Desert Storm—now found themselves flying daily missions over Iraq's cities and deserts. Their presence not only expanded the pool of available aircraft, but also underscored the reality that the burden of OIF would be shared across the entire Air Force.

Understandably, the nature of air power during this new phase carried its own challenges. Insurgents had no radar-guided missiles, no integrated air defense system, and no fighters to challenge coalition control of the skies. Yet they adapted in their own way, using urban terrain to mask movements, exploiting crowds as human shields, and dispersing across the dense cityscapes. For the F-16 pilots, this meant a new level of precision targeting. A bomb delivered during a high-tempo CAS mission had to strike its target without harming nearby civilians. The fighter that had once raced through MEZs now circled patiently overhead—watching, waiting, and delivering its weapons only when the parameters were clear.

Still, the sortie counts remained high. From 2003 onward, coalition air wings flew thousands of missions every year, the vast majority of them F-16s. Every squadron deployed for months at a time. Their pilots would fly daily patrols, and their maintenance crews worked around the clock to keep the aircraft mission-ready. The demands of such an enduring conflict would push the Air Force's personnel system to its limits. Yet the rotations continued, sustained by their proven expeditionary model.

By 2004, as the insurgency grew, so too did the coalition's reliance on air power. The various urban battles always saw the F-16 overhead, delivering strikes that shattered strongpoints and tipped the scales in favor of Allied ground forces. For the pilots, however, counterinsurgency demanded a different mindset. Their missions were marked by long hours of loitering and the constant challenge of discriminating targets in complex terrain.

In fact, among the shifting sands of Iraq's insurgency,

few battles would carry the same weight, ferocity, or collective risks as the Battle of Fallujah.

Nestled along the banks of the Euphrates, Fallujah had become a hotbed of insurgent activity after the fall of Saddam Hussein. In fact, by 2004, local insurgents had transformed the once-peaceful municipality into a fortress of resistance, carving strongholds into the city's mosques, schools, and homes. To reclaim it, coalition ground forces would have to fight in the streets. Meanwhile, the F-16 would reign over the city skies, turning the tides of urban combat through the power of tactical air support.

The Viper's victory over Fallujah began not with a sweeping bombardment, but with a quiet revolution in precision firepower. In September 2004, an American F-16 carried a brand-new weapon into combat: The 500-pound GBU-38—the latest incarnation of the JDAM. Unlike the heavier, unguided bombs that carried the risk of collateral damage, the GBU-38 promised surgical precision within the narrow confines of the urban battlespace.

Major Brian Wolf, then flying with the 160th Fighter Squadron, remembered the mission clearly. On the morning of September 12, his jet was loaded to capacity: "We got in-country and the loadout we were carrying was GBU-38s and slant loads of GBU-12s. We also had two AMRAAMs, 500 rounds of 20mm and the LITENING II pod." At first, the mission seemed routine—pipeline patrols and surveilling vital infrastructure. But as daylight broke over the desert, new coordinates came through. There was a suspected insurgent stronghold in Fallujah.

Wolf and his wingman turned their targeting pods toward the city. "We were looking at the target complex with our LITENING II pods," he recalled. "We could see the target from more than 20 miles away."

Then came the pivotal moment.

A pair of GBU-38s rained down from above, guided by satellite precision, slamming into the target with devastating accuracy. Wolf later recalled: "It was a distinct and exact impact, and the bombs had hit within a second of each other." In the blink of an eye, insurgents who had believed themselves to be well-hidden, suddenly perished in the flames of angry JDAMs.

The assessment that followed was surreal.

Circling back, Wolf and his wingman watched the aftermath from the air. The GBU-38 had obliterated the so-called "safe house." In fact, the explosion had been so massive, other coalition aircraft swarmed into orbit, "like sharks converging for the kill," he recalled. All were eager to see the effects of the new JDAMs in combat. Back at Al Udeid, Wolf sat in the mess hall as the strike footage replayed on the television screen. From the infrared reel, he watched the insurgents step into view. "One guy took a drag off of a cigarette," he said, "and you could see the heat from it. As he took it out of his mouth, the GBU-38 impacted and the war ended for him."

That single strike carried profound meaning. It was the combat debut of the GBU-38, a weapon that would define the battles of Fallujah and beyond. More than technology, it represented a shift in the way air power would be wielded on the urban landscape. It was precise and restrained, but utterly decisive when unleashed.

As the battles for Fallujah unfolded later in 2004, the F-16s continued to shape the fight. The city soon descended into a grinding battle of attrition, where Marines fought block by block and the Vipers brought death from above.

Wolf's squadron mates echoed the essence of their

mission: Intense CAS, presence patrols, and a lot of patience. It spoke to the multi-faceted role of air power in the city. Sometimes the Vipers flattened buildings being used as firing points, bringing fire and steel against the entrenched jihadis. But more often than not, they roared in low over the city, using their speed and noise to scatter enemies before they could fire. Indeed, the roar of afterburners through narrow streets was often enough to deter an ambush. And if it wasn't, a precision-guided death warrant soon followed.

Still, every sortie demanded the utmost discipline. In the crowded neighborhoods of Fallujah, mistakes could cost civilian lives and fuel the insurgency. All along the urban battlefront, F-16 targeting pods scanned rooftops, alleyways, and city roads, relaying intelligence to ground units and waiting for the call to strike. When insurgents fortified houses or engaged Marines with suppressive fire, the response from above came within minutes.

All told, Fallujah demanded more from the F-16 than speed or power. It demanded patience, vigilance, and quick adaptability. The Viper, once built for Cold War dogfights and high-speed strikes, had proven its worth in the tightest battlespace of the modern age. Just as they were doing on the unconventional battlefields of Afghanistan; the F-16s were influencing the tempo of Iraq's counterinsurgency.

When the fighting subsided and the Marines declared victory over Fallujah, the contribution of the F-16 was measured both by the number of targets it destroyed *and* the number of lives it saved. The first combat use of the GBU-38 signaled a new era of precision-guided warfare; and the months that followed proved the indispensability of air power to urban combat.

The recollections from pilots like Brian Wolf preserve that reality. From the thunder of bombs to the roar of afterburners, the Viper's presence over Fallujah was constant. The insurgents never knew when the sky would erupt; and the Marines never doubted that help would come. Thus, the story of Fallujah belongs not only to those who fought in its streets, but also to those who circled the skies above.

By 2006, the war in Iraq had entered a harsher season. Sectarian violence spiked, the insurgency adapted, and the urban battlespace demanded an increasingly surgical approach. Yet in the growing crucible of urban combat and counterinsurgency, the F-16 remained a constant companion above coalition ground forces. During this time, the 332nd AEW (having relocated to Balad Air Base) continued patrolling the skies over Iraq, their Vipers switching from CAS to reconnaissance to overwatch by the hour.

As it turned out, one such call arrived on the evening of June 7, 2006.

From the earliest days of the insurgency, Abu Musab al-Zarqawi had emerged as a leading figure among the foreign militants in Iraq. The mastermind behind a string of deadly bombings and beheadings, al-Zarqawi had become the fiery node of Al-Qaeda in Iraq (AQI)—the Mesopotamian branch of bin Laden's terror network. On this sweltering June day, Allied intelligence pinpointed him inside a safehouse near Hibhib, just north of Baqubah.

And a pair of F-16s from the 332nd were tasked to terminate him.

Thus, in the black geometry of their final attack run, the lead Viper released its GBU-12 laser-guided bomb, followed a split-second later by a GBU-38 JDAM. Both 500-pound weapons struck with near-simultaneous fury. Following a brilliant flash of molten bricks and mortar, the house collapsed into itself. Zarqawi and several of his associates were killed. The airstrike's fatal arithmetic (two bombs, one safehouse) demonstrated the vitality of air power as a tool for modern-day manhunts.

The aftermath confirmed what the weapons had promised: A clean, decisive blow with unequivocal results. It was a moment when patience, diligent tracking, and airborne precision converged to eliminate AQI's notorious leader. Archival footage released by the US Department of Defense showed the impacts of the GBU-12 and GBU-38, grainy yet unmistakable, as the target site vanished into a single blooming detonation.

Across Iraq, the operational story behind that single strike played out many more times in lesser-known forms. The F-16s flew protective patrols over infrastructure, scanned highways with their targeting pods, and vectored to wherever troops needed tactical air support. In many ways, their OIF missions reflected what was already happening in Afghanistan: F-16s watching, waiting, and striking when the geometry was right or when the ground scheme demanded it.

Five months after the victory of eliminating al-Zarqawi, however, another F-16 story wrote itself in much harsher ink. On November 27, 2006, Major Troy Gilbert—a veteran Viper pilot and the 332nd Expeditionary Operations Group's chief of standardization—took flight on a CAS mission. The sortie began as an NTISR flight, the kind of long-orbit vigilance that had come to define the 2006 deployment season.

Mid-sortie, however, a coalition helicopter made a crash landing near Fallujah. The ground troops moving to secure the crash site then came under attack from small arms fire, RPGs, and civilian vehicles with mounted weapons. Gilbert and his wingman answered the call for CAS. But with his wingman running low on fuel, Gilbert split the formation, ordering his wingman to vector off towards the nearest tanker. Gilbert, meanwhile, stayed on station with the JTAC, ready to line up his 20mm autocannon for a strafing attack.

Having positively identified hostile vehicles, Gilbert tore the F-16 into a high-angle strafe at close range, firing off a burst of 20mm rounds that bounded up to the insurgents' truck, quickly scattering the enemy's charge. Recovering at roughly 200 feet above the desert floor, he turned again to shield the friendly troops. But in the steepening press of that second attack—amidst the dust, the speed, and the narrowing cone of a strafing run—his aircraft descended below the threshold for recovery and struck the ground with a fiery explosion.

Gilbert was killed on impact.

The Air Force's board of inquiry determined that Gilbert's "channelized attention" and target fixation in the dynamic, high-stress environment had been the contributing factor to his demise. In other words, it was an aviator's fierce determination to protect his ground-based comrades, but taken to an extreme and ragged edge.

A Marine Quick Reaction Force (QRF) then had to fight to secure both crash sites—first the helicopter, then the F-16. But in the ensuing chaos, enemy insurgents got to the F-16 first and removed Gilbert's body before the QRF could reach him. That theft of Gilbert's body (and the denial of his dignity) was the opening chapter to a long and bitter search. It would become a tragedy marked by ten years of pursuit, negotiation, detective work, and the public heartache of a family waiting for the return of their beloved pilot.

The first fragments of recovery, however, came rather quickly. DNA recovered from the crash site allowed Gilbert's partial remains to be buried at Arlington in December 2006.

Then, the years dragged on.

In 2012, additional remains were recovered, allowing for a second funeral in December 2013. It was a second reprieve, but not yet an ending. Gilbert's story became a covenant, of sorts: The promise of a search with no expiration date.

Finally, in the summer of 2016, an Iraqi tribal leader approached a US military adviser near al-Taqaddum. He held what he claimed were the remains and flight gear of an American pilot lost during the war. Forensic confirmation followed in September at Dover AFB. And, in early October, Gilbert's final remains arrived in the US under a dignified transfer. On December 19, 2016—ten years and three funerals after the crash—Major Troy Gilbert was laid to final rest at Arlington.

From Hibhib to the outskirts of Baghdad, the 2006 season distilled what the F-16 could bring to the counterinsurgency fight. The Zarqawi strike showed a multirole fighter able to pair laser-guided accuracy with GPS-guided certainty—one GBU-12 for the terminal cue, one GBU-38 for the assured finish. It was an economy of force that amplified the coalition's wider campaign. And the Gilbert mission, tragically, highlighted the occasional cost to keep that economy of force intact at the ground level. During these heated CAS missions along the urban battlefront, a pilot would often fly lower and closer to the ground because seconds mattered and American lives were in the balance.

In between, countless orbits and taskings stitched the urban battlefield together. Vipers from the 332nd often teamed with A-10s, helicopters, MC-130s, and Predator drones. Whether operating alone or in tandem with other air assets, these F-16s kept a tight liaison with JTACs and QRFs—warning, watching, and striking when needed. The US Air Force's official description of the 332nd's mission set reads like a ledger of 2006 itself: CAS, NTISR, precision strikes, and presence patrols. It was a year when the Viper's Cold War lineage met the hard arithmetic of urban warfare, and proved equally suited to both.

From Surge to Sunset (2007-11)

By early 2007, Iraq stood on the brink of collapse. Sectarian violence and a resilient insurgency had seized the country in a storm of bombings, ambushes, and targeted murders. Into this tempest came the American "Troop Surge." In January 2007, President Bush ordered an additional 20,000 troops into Iraq to secure Baghdad and the Anbar Province, while extending the deployments of several Army and Marine Corps units already in country. All told, it was a gamble that more "boots on the ground," paired with relentless air power, could turn the tide of the counterinsurgency.

Above the convoys, city streets, and villages, the F-16 continued to carve its role as both guardian and avenger, maintaining its vigil in the skies over Baghdad, Basrah, Baqubah, and beyond. From 2007 until the final drawdown in 2011, the F-16 evolved from a strike bomber to a sentinel of stability, helping to facilitate the dawn of the new Iraqi Air Force.

When the Troop Surge began in the spring of 2007, the F-16s delivered near-constant support. On February 11, for example, a sudden firefight erupted in the streets of Latifiya. That day, an explosive ordnance disposal team was inching forward to neutralize an IED. But as the technicians began to defuse the roadside bomb, a squad of local insurgents opened fire. The fight might have turned deadly, had it not been for the F-16s orbiting overhead. Their arrival broke the ambush before it escalated, silencing the insurgents' gunfire with the thunder of jet engines.

Eight days later, on February 19, insurgents struck again, detonating an IED near Tikrit. Coalition convoys were vulnerable, exposed on a road flanked on both sides by potential killers. But once more, the sight of Vipers descending from the sky changed the equation. As the F-16s swept low over the area, the enemy promptly ceased fire. Elsewhere that day, pilots reported groups of insurgents gathering in alleys, feeding this intelligence to ground controllers who tightened their cordon. By nightfall, coalition aircraft had flown more than fifty CAS missions across Iraq, an early testament to the rhythm of the Troop Surge.

As spring turned to summer, the surge began to reach its crescendo. And on June 30, the city of Baqubah became a proving ground of sorts. Insurgents had been turning homes into arsenals, stockpiling all varieties of ordnance to use against coalition forces. But a two-ship flight of F-16s, guided by local JTACs, released their GBU-12 laser-guided bombs onto one such residence, collapsing it into dust. Elsewhere, they safeguarded troops as they cleared the city block-by-block, swooping overhead as squads moved from house to house. Meanwhile, in Al Miqdadiyah, they provided overwatch for a bombed Iraqi Police station, making sure insurgents could not regroup there. Over Taji, they executed more shows of force—ear-splitting, high-speed passes that scattered enemy fighters without a shot being fired.

Throughout July, the tempo didn't diminish. From Balad Air Base, F-16 crews launched in tight rotations, each sortie a mission of vigilance and violence. On July 28, their efforts spanned Baghdad and beyond, with shows of force and precision strikes designed not only to destroy but to intimidate, severing the will of insurgents who watched the skies knowing reprisal could come at any moment. The following week, on August 4, a flight of F-16s destroyed several dirt bridges being used by insurgents in Baghdad. These were not grandiose targets—no bunkers, arsenals, or safe houses—but crude crossings used to shuttle militants and their weapons into battle. As simple as these infrastructure targets may have been, their destruction carried a disproportionate weight. Indeed, it had severed the insurgents' immediate lifelines and isolated the respective enemy cells. In the calculus of counterinsurgency, such strikes were not merely tactical; they were surgical blows to the enemy's ability to sustain its fight.

As 2008 opened, the air war continued to press insurgents on every front. On January 6, American Vipers over Baghdad struck a number of houses seeded with IEDs, their GBU-38 JDAMs collapsing walls that had concealed death for passing convoys. In Babil, they leveled hostile compounds, while in Nasiriyah, they fired their 20mm cannons and GBU-12s to liquidate teams of entrenched gunmen. Every CAS mission often concluded with a show of force, ripping low through the sky to warn that further attacks were futile.

On January 9, a flight of F-16s joined a Navy F/A-18 patrol in hammering Al-Qaeda positions outside Baqubah. Their weapons *literally* reshaped the battlefield: GBU-12s caved in the enemy's structures and the GBU-38s cratered roads to deny insurgent passage. In Tikrit, the simple roar of a Viper executing a show of force was enough to disperse enemy fighters. By day's end, coalition aircraft had flown sixty-two CAS missions, proof that the aerial hammer remained tireless. And on January 15, the pattern repeated itself. In Basrah, F-16s dropped GBU-38s on an IED cache, silencing one more enemy threat. Other flights struck hostile positions in Babil and Al Kut, while others again swept low in deterrent passes.

From there, the pressure did not relent. On January 21, F-16s returned to Baghdad, destroying houses booby-trapped with explosives, then fanning out across Basrah to conduct repeated shows of force. One week later, another airstrike underscored the agility of the mission: F-16s destroyed a vehicle carrying insurgents in Baqubah and neutralized a rocket launcher near Basrah with a GBU-38.

Taken together, the missions of 2007-2008 revealed the layered utility of the F-16 in counterinsurgency. From the collective missions in Iraq and Afghanistan, it was clear that the Viper was no longer the Cold War dogfighter it was designed to be. In the skies over this two-front war, it was a hovering guardian, a laser-focused instrument of deadly air support, and at times, a psychological weapon.

By the close of 2008, the Troop Surge had checked the insurgency's momentum. While the peace was fragile, the insurgents had learned that no ground was safe under the gaze of the American Vipers. For two years, the skies above Iraq had been etched with contrails and the echoes of afterburners, marking the presence of a fighter jet that had transcended its origins.

As US ground forces prepared to withdraw, the focus shifted from kinetic operations to mentorship and building long-term self-sustainability for Iraqi forces. To that end, the US Air Force established the 52nd Expeditionary Flying Training Squadron in March 2007 at Kirkuk Air Base. It was to be a fixed-wing flight school, using prop-driven Cessna planes as their entry-level trainers. It was a critical step for the emergence of a newly-independent and revitalized Iraqi Air Force. By October 2008, the 52nd had graduated its first class of Iraqi fixed-wing pilots.

Simultaneously, US military support transitioned to an advisory role. By mid-2010, the last US combat brigade exited Iraq, bringing an end to Operation Iraqi Freedom. From there, the mission became Operation New Dawn, with a clear directive of training and advising Iraqi forces. While ground combat receded, the air domain remained central: F-16s continued to provide overwatch, surveillance, and tactical support when requested. But now, Iraqi pilots were increasingly flying alongside their US counterparts, building the institutional knowledge necessary to stand on their own.

The final phase of American combat presence unfolded steadily. Even as the last ground troops departed in December 2011, the US continued training Iraqi airmen and developing their air force's infrastructure. Baghdad signed lucrative contracts for eighteen *Block 52* F-16C fighters, placing Iraq on the road to air sovereignty. The sales contract also included pilot and maintenance training, reinforcing the new strategic partnership between the Iraqi and American Air Forces.

Weeks later, a sale of eighteen additional F-16s was approved, signaling that the Iraqi Air Force would field up to three dozen Vipers by 2018. These jets came equipped with the AIM-9 Sidewinder and AIM-7 Sparrow missiles, targeting pods, and conformal fuel tanks—an aircraft package befitting a sovereign nation's air force.

As the final US air bases began transferring control, Balad Air Base returned to Iraqi command on November 8, 2011. Though the American Vipers had departed, the flame of F-16 operations remained alive in the hands of Iraqi airmen. Balad would later become home to the Iraqi Air Force's 9th Fighter Squadron, operating the *Block 52* F-16s. Despite security and political challenges, the delivery of these F-16s represented more than just modern hardware. It was the sign of an enduring alliance, a guarantee of strategic autonomy, and a continuation of American influence through cooperative air power.

From the Troop Surge of 2007 to the final withdrawal in 2011, the F-16 remained constant in Iraq's skies. In many ways, it embodied the spirit of transition—from supporting American boots on the ground, to Iraqi pilots raising their own wings over Baghdad.

Every strafing run, every JDAM engagement, every pilot trained in Kirkuk, and every contract signed for Iraqi Vipers represented a step toward Iraqi self-sufficiency and institutional maturity. When the last coalition airman departed, the story of the F-16 in Iraq had evolved from combat dominance to a legacy of empowerment.

When the war began in March 2003, the F-16 was among

the first to pierce the skies over Iraq. From desert runways in Saudi Arabia, Qatar, and Kuwait, American Vipers thundered into the darkness to deliver the opening blows of the invasion. Just as they had done in the former Yugoslavia and over the Hindu Kush, F-16s had once again proven themselves as the multi-role strike aircraft of choice. With its unparalleled adaptability, the Fighting Falcon helped shape the tempo of the campaign that toppled the regime of Saddam Hussein.

Yet the fall of Baghdad was not the end; it was the beginning of a far longer struggle. As coalition forces found themselves in the labyrinth of insurgency, the F-16 remained overhead, evolving to meet the war's shifting demands. It became a guardian to American convoys threading the roads of Iraq; a watchful eye over troops patrolling restive neighborhoods; and a tool of deadly precision when insurgents hid within homes, mosques, or schools. With targeting pods and JDAMs, Vipers became the embodiment of counterinsurgency air power.

During the Troop Surge of 2007-08, its wings rarely left the skies. From Baghdad to Basrah, the F-16 shattered enemy compounds, collapsed IED caches, and carried out shows of force that scattered fighters before the battle could even begin. During those years, the roar of its engines and the glint of its gray wings became as much a part of Iraq's landscape as the armored columns below. To American troops pinned down within alleys or along the desert highways, the sight of an F-16 was a promise from above: *You are not alone.*

The latter years of the war brought a new phase. As US combat forces prepared to depart, the mission shifted from domination to mentorship. F-16s still flew in support of counterinsurgency operations, but their presence increasingly symbolized the forthcoming handover to the Iraqi Air Force. At Kirkuk and Balad, American instructors trained Iraqi pilots, guiding a new generation toward mastery of the very aircraft that had once loomed over them in battle.

By December 2011, the long vigil had ended. The last American combat forces departed, and with them the F-16s that had once thundered over the valleys of the Tigris and Euphrates. In their place, Iraqi pilots shouldered the responsibilities of sovereign air power, inheriting not just the aircraft, but a legacy of partnership and sacrifice. The F-16's story in Iraq was no longer just an American tale. It had become an Iraqi one as well.

An American F-16 crew performs its post-flight checks at Aviano Air Base, Italy, after the pilot returns from a mission in support of Operation Odyssey Dawn, March 2011. Odyssey Dawn was part of the larger international response to the civil war in Libya, which resulted in the collapse of Muammar al-Gaddafi's regime. *US Department of Defense*

Chapter 7:
Arab Winter

Throughout 2011, the hopeful cries of an "Arab Spring" echoed across North Africa and the Middle East, promising a new era of reform, liberty, and self-determination. Yet, in the years that followed, the region instead devolved into a vicious cycle of chaos and violence that resembled anything but Spring. Popular uprisings toppled dictators, but in their place came civil wars, jihadist insurgencies, and foreign interventions. The geopolitical vacuum was filled not with peace, but with fire.

Against this backdrop, the F-16 once again became a frontline player for both the Arab and Western air forces. From the Mediterranean to the edge of the Arabian Peninsula, the mighty Viper would bear witness to—and play a decisive role in—conflicts that reshaped the balance of power throughout the Middle East. In Libya, NATO's intervention to assist the rebels against Muammar al-Gaddafi saw European, American, and Arab F-16 squadrons enforcing No-Fly Zones and striking targets in an air campaign that ended the Libyan strongman's rule.

Meanwhile, a new enemy emerged in Iraq and Syria. The rise of the so-called "Islamic State of Iraq & Syria" or "Islamic State of Iraq & the Levant" (ISIS/ISIL) thrust the F-16 into another long war, as Arab partners and coalition air forces joined in Operation Inherent Resolve. Vipers struck convoys, bombed training camps, and destroyed urban strongholds from Mosul to Raqqa, becoming a constant presence in the skies of a conflict that lasted longer than many had foreseen.

Farther south, at the edge of the Arabian Peninsula, the F-16 became the workhorse of the Saudi-led coalition in the Yemeni Civil War. Emirati, Bahraini, Jordanian, and Moroccan Falcons flew sorties against Houthi rebels in a grinding war marked by shifting frontlines, humanitarian catastrophe, and international controversy. For many of these air forces, the Yemeni Civil War was the most sustained combat use of their F-16 fleets to date.

Taken together, these three conflicts formed the trilogy of an "Arab Winter," each different in cause and character, yet unified in their reliance on the F-16 as both a tactical weapon and a sign of geopolitical solidarity. The Falcon's presence not only underscored its enduring relevance four decades after its maiden flight, but also the centrality of air power in a region perpetually caught in the crossfire of armed conflict.

Libya (2011)

The Libyan Civil War of 2011 erupted in the space between high ideals and hard constraints. In February and March, protests inspired by the Arab Spring were met with violent repression from the Gaddafi regime. As Gaddafi loyalists pushed eastward into Benghazi, the United Nations passed two resolutions in rapid succession: Resolution 1970, which imposed economic sanctions and referred Libya to the International Criminal Court; and Resolution 1973, which authorized "all necessary measures" to protect civilians, including an arms embargo and a No-Fly Zone.

On March 19, a coalition led initially by the US, UK, and France opened with air and missile strikes to suppress Libyan air defenses and halt Gaddafi's armored forces on their approach to Benghazi. The US contribution during that first phase—Operation Odyssey Dawn—included the F-16CJ Wild Weasels alongside other assets. Within days, NATO assumed unified command, thus re-branding the campaign Operation Unified Protector (OUP), which ran from March 31 to October 31. By the end, NATO had flown nearly 26,500 sorties. It was an air war fought to a decisive political end with no Allied combat losses.

For the F-16 community, Libya once again showcased the aircraft's versatility across international fleets. Much of the lineup and task delegations were reminiscent of the latter-day mission in Yugoslavia. For example, American Vipers took the lead for strike missions and suppressing enemy air defenses; Nordic and Belgian forces took the reins for precision attacks; while the Dutchmen handled air policing and ISR support.

This time, however, the European task force was joined by a Gulf contingent that carried heavy political weight. The Viper's ubiquity meant it wasn't just a participant; it was a key player in supporting the UN mandate and restricting Gaddafi's ability to concentrate firepower and maneuver.

The coalition's sequencing reflected the mandates of Resolution 1973. First, they would neutralize Libyan air defenses. Second, they would prevent loyalist air-to-air operations. Third, coalition jets would interdict ground forces threatening populated areas. Finally, they would hold this protective umbrella until the regime forces had collapsed. From their command center in Italy, NATO built a combined air tasking cycle that coordinated multinational packages, standardized targeting procedures, and synchronized airborne surveillance.

Within this framework, F-16 detachments fell into roles

based on national caveats and capabilities. For instance, some nations cleared their jets for deliberate airstrikes, while others authorized only defensive counter-air or reconnaissance support. A few key choices, like the Netherlands' decision to emphasize No-Fly Zone enforcement and reconnaissance flights, shaped which F-16s did the heaviest lifting and when.

In the opening days of Odyssey Dawn, the US Air Force sent their F-16CJs to attack Libyan radar and SAM sites. Eight F-16CJs (paired with four F-15E Strike Eagles) launched in the first wave—the spearhead of a classic SEAD campaign. This, in turn, would facilitate the low-altitude bombing runs perpetrated by other coalition-member F-16s. All told, these inaugural SEAD missions showcased the American Viper in its most specialized role: Soaking up radar emissions, cueing HARMs, and smashing enemy radar grids while cruise missiles and other strike aircraft engaged critical targets.

As NATO took over (and SEAD demands tapered off), American F-16s continued to fly presence patrols, escort, and strike missions as needed. But Washington soon receded into an enabling posture (taking on surveillance and reconnaissance flights, providing aerial refuels, etc.) so that European and regional partners could carry the kinetic day-to-day operations. That political choice made the performance of the European F-16s even more central to OUP's outcome.

In a twist of geopolitical irony, Denmark's F-16s became the standard-bearers for precision airstrikes in Libya. The Royal Danish Air Force (RDAF) deployed six F-16AMs to Sigonella Air Base, Sicily, and proceeded to fly for the remainder of the conflict. Official Danish records confirm nearly 600 missions and 923 precision-guided bombs delivered—remarkable outputs for a six-jet task force. Indeed, it was a clear indication of how NATO's smaller air forces could deliver strategic weight. Public data and news tallies at the time noted that Denmark's percentage of ordnance released was among the highest relative to its size.

The RDAF playbook relied heavily on the Viper's targeting pods and GPS/LGB inventory to liquidate enemy ground forces, often under dynamic targeting rules that required fast engagement cycles with strict collateral-damage procedures. The long duration of their deployment forced Denmark to manage crews and spares carefully: Ramping up to four sorties per day during the "hot phase," then throttling back as target density fell, all without impacting readiness. But these mission cycles also highlighted a familiar constraint that had plagued European air forces for decades: their limited stockpiles. Indeed, by mid-campaign, the Allies were struggling to replenish their ammunition stores—a logistical lesson that would influence NATO's future planning cycles.

Six Norwegian F-16AMs joined the fight from their forward base in Souda Bay (Crete). Oslo had authorized strike operations from the outset and, within the first few months, the Royal Norwegian Air Force had become one of the highest-tempo contributors of the campaign. By the time Norway rotated out in July 2011, their F-16 contingent had flown more than 600 sorties and dropped nearly as many bombs—a significant share of the coalition's early kinetic effort.

The Belgian Air Force, meanwhile, deployed six F-16AMs to Araxos (Greece). Although runway issues initially restricted operations, Belgian Vipers soon settled into a notable niche: Nighttime strike packages that hunted vehicles, hardened shelters, and supply depots. Under these auspices, the Belgian detachment reported a total of 448 missions with 365 bombs dropped. While the exact figures vary among sources, the narrative is consistent: Belgium's Air Force brought a disciplined, high-yield F-16 contribution that grew more specialized as the war in Libya matured.

Belgium's night raids exemplified another advantage of the F-16 in multinational air wars: Common tactics and systems maturity shorten the time from deployment to effects. Belgian, Danish, Dutch, and Norwegian F-16 fleets had trained together for years under the EPAF banner; and the war in Libya had turned that standardized playbook into a nightly production.

The Royal Netherlands Air Force (RNLAF) positioned six F-16AMs at Decimomannu (Sardinia) with a KDC-10 tanker. The Dutch government had initially restricted their F-16s to No-Fly Zone enforcement, arms embargo support, and ISR missions. Throughout the campaign, the RNLAF logged 639 sorties and nearly 3,000 flight hours, feeding target-development data into NATO's planning cycle and providing air patrols that freed up other F-16s for deliberate attack. Although The Hague kept its fighter jets on a tight leash, the operational point remains: In a complex mandate, "non-kinetic" F-16 roles (ISR missions, air policing, etc.), were an indispensable part of the air power equation.

Although the Kingdom of Jordan volunteered its Air Force to safeguard humanitarian air corridors and provide "logistical support," its decision to send F-16s into a NATO-run enterprise was both monumental and symbolic. Indeed, Arab F-16s had aligned with a UN-backed mission at a time when regional legitimacy mattered. Amman's Foreign Ministry announced the deployment in April 2011, noting their intent to maintain a primarily non-combatant role. The Jordanian mission, therefore, was a reminder that "presence" can be strategic currency in coalition air campaigns, even when rules of engagement stop short of kinetic airstrikes.

Another Arab partner, the UAE, sent six F-16E/F *Block 60* "Desert Falcons" (along with six Mirage fighters) to Decimomannu—a first-tier Arab air arm flying one of the most high-tech F-16s then in service. The UAE's participation was as politically potent as their Jordanian counterparts: A non-NATO, Arab League member was contributing to a UN-mandated operation in Libya. While the UAE's Mirages drew early headlines, the deployment of the Desert Falcons—equipped with AESA radar, conformal tanks, and advanced avionics—signaled a generational shift in the Gulf States' capacity for air power.

After the SEAD opening, OUP settled into a high-tempo battle rhythm: Armed overwatch near threatened cities, pinpoint airstrikes against Command & Control facilities, and interdiction of loyalist armor and artillery. F-16s—especially from the EPAF fleets—excelled at rapid re-tasking: Pod-driven searches, tight liaisons with FACs and JTACs, and delivering effects with quick turnaround times. Standard loadouts included the GBU-12/16/24s, JDAM variants, and 20mm ammunition. These loadouts gave the pilots a near-unprecedented menu depth. And because the ground picture in Libya was constantly changing, precision and persistence mattered just as much as raw sortie count.

Danish and Norwegian F-16s often drew the dirtiest work during the early months—engaging enemy armor and artillery positions—while Belgian night strikes increased throughput against critical regime targets. Dutch F-16s helped keep the airspace sanitized and the intel picture fresh, while Jordanian and Emirati F-16s provided escorts along with humanitarian overwatch.

By the summer of 2011, Gaddafi's forces found it increasingly difficult to mass or move. The "No-Drive Zone" effect (never formally declared, but functionally enforced) grew from dozens of daily interdictions that denied loyalist ground forces the ability to sustain their offensives. OUP's culminating act was, essentially, an air-enabled ground maneuver. As rebel forces advanced on Tripoli and Sirte, the air campaign continued to peel away at Gaddafi's military. NATO declared the mission complete on October 31, 2011, mere days after Gaddafi's downfall.

Yet the campaign's clean end from 15,000 feet belied a much bigger problem on the ground. OUP was intended to facilitate regime change, not stabilize Libya. And the postwar vacuum became a cautionary tale about the use of coercive air power. Still, for the F-16 community, Libya had been an operational success. Multirole fighters, across a multinational fleet, under a tightly-efficient NATO command, had delivered sustained effects at coalition scale.

From an operational standpoint, Libya taught the air power community a number of valid lessons. First, SEAD operations still paved the way for air superiority, and the F-16CJ was still the fastest trailblazer in the West. Libya's air defenses were dated, but they were not trivial. The opening salvo of HARMs-shooting F-16s, paired with cruise missiles and multi-layered reconnaissance, underscored that "first-night" operations were both an art and a science. Moreover, it testified that the Viper's Wild Weasel lineage was still relevant wherever an air coalition needed to fly low and often.

Second, Libya showed that smaller Allied fleets could punch above their weight class, albeit briefly. Danish and Norwegian output was extraordinary, but both were straddled by finite stockpiles and force-generation limits. The Libyan Civil War thus forced NATO to re-examine their common ammunition pools, common operating procedures, as well as their surge logistics.

Third, the defense community saw how non-kinetic roles became force multipliers. Dutch F-16s showed how fighter patrols, recon patrols, and tanker escorts freed other aircraft for critical air-to-ground taskings—and how political constraints can be operational assets when they're aligned to mission needs.

Fourth, OUP showed how regional politics traveled on the wings of air power. The UAE and Jordan flew for more than targets; they flew for legitimacy. An Arab F-16 presence inside a UN/NATO command structure mattered for maintaining the mandate's regional credibility—a lesson later applied against ISIS. Finally, judging from the aftermath of the conflict, the rebels' proved that while air power can *protect*, it cannot *rebuild*. The F-16 had done its job. But stabilization was another problem entirely.

Operation Inherent Resolve

Meanwhile, in the Syrian theater of the Arab Spring, similar cries of hope and defiance had likewise devolved into a civil war. Protests against the reign of Bashar al-Assad were met with the same brand of brutality seen in Libya. Government forces fired on demonstrators, and entire cities rose in open revolt. What followed was a descent into chaos, as defectors from the Syrian Army formed the Free Syrian Army and insurgent groups multiplied across the fractured landscape. By 2012, it was clear that Syria wasn't just another Arab civil war; it had become a battlefield of competing factions—loyalists, secular opposition, Islamist brigades, and Kurdish militias—each carving out their own territory amid the ruins.

Naturally, the power vacuum became a magnet for extremists. From the remnants of al-Qaeda in Iraq (AQI), a new organization emerged: the Islamic State of Iraq and Syria (ISIS). Born in the deserts of al-Anbar and hardened by years of insurgency against US forces in Iraq, ISIS crossed into Syria and quickly exploited the chaos. They seized towns in the east, including Raqqa, and established a proto-state with its own courts, tax laws, and brutal codes of justice. Their message was clear: While others fought for influence, ISIS claimed divine authority to rule.

In June 2014, ISIS made headlines sweeping across northern Iraq and capturing the city of Mosul. Iraqi Army units melted away in panic, abandoning their US-supplied weapons and vehicles, which ISIS eagerly absorbed into its arsenal. Days later, Abu Bakr al-Baghdadi ascended to the pulpit of Mosul's al-Nuri Mosque, declaring the creation of a new caliphate, and calling upon Muslims everywhere to join his cause.

Over the next few years, the self-proclaimed caliphate would become infamous for its atrocities: Public beheadings of hostages; the massacre of Yazidis on Mount Sinjar; and campaigns of cultural destruction against ancient heritage sites. Propaganda videos flooded the internet, designed to terrify adversaries and inspire recruits. Thousands of foreign fighters answered the call, streaming into Syria and Iraq to wage jihad under the black flag of the Islamic State. At its height, ISIS controlled territory roughly the size of Great Britain, stretching from the suburbs of Aleppo to the gates of Baghdad.

The reign of terror quickly grew to a size and intensity that the international community could no longer ignore. Iraq teetered on the brink of collapse, and Syria was already going down in flames. The rise of ISIS was no longer just a regional threat; it was now a global security risk. In August 2014, US airstrikes began in northern Iraq to halt ISIS's advance toward Erbil and to relieve the besieged Yazidis. Within weeks, President Obama announced the formation of a multinational coalition. More than 70 countries pledged support, from NATO allies to Arab partners, all united in their aim to destroy ISIS.

Thus began Operation Inherent Resolve (OIR).
The fight against ISIS would be a long, grinding campaign—fought in the skies over Iraq and Syria, in the alleys of Mosul and Raqqa, and in the hearts of millions who lived captive under the black flag of the Islamic State. Into this storm came the F-16, now tasked with breaking the caliphate's grip and ensuring that ISIS met its ultimate demise.

US Operations

OIR formally began on August 8, 2014, with the stated objective to "degrade and ultimately destroy" ISIS through sustained air strikes. The new Combined Joint Task Force–OIR took the fight to ISIS primarily by air, with F-16s joining established bomber and fighter units in the skies over Iraq and Syria.

Early on, however, ROE and logistical constraints hampered their effectiveness. As one pilot later recalled: "During each sortie during the campaign's first month, we would watch all sorts of [IS]-related activity going on in Syria. The targets were definitely out there for us to kill. I saw them day after day. We flew over such targets in Syria day in and day out with bombs on our jets...and still we did nothing about it." He lamented that with more aggressive targeting of ISIS's leadership and oil infrastructure from the outset, "we could've done some serious damage and saved lots of lives." This hindsight underscores how throughout late 2014, America's F-16s soared overhead while ISIS slipped beneath the radar.

Nevertheless, American Viper pilots flew combat missions almost every night. And although no American F-16s were lost to enemy fire during the first year of OIR, there was a tragic crash that underscored the grit and raw stakes of expeditionary air power. On November 30, 2014, Captain William DuBois of the 77th Fighter Squadron, crashed during takeoff just as dawn was breaking over his unit's air field in Jordan. He was killed instantly. Dubois had been in-country for only six weeks, yet he had already flown eighteen combat missions, striking ISIS targets throughout Iraq and Syria.

And just as they had done during Operations Enduring

Freedom, Iraqi Freedom, and the No-Fly Zone Patrols, these F-16s never flew without the support of AWACs and aerial tankers nearby. In fact, on one occasion, a KC-135 tanker crew found themselves as the unlikely savior for an ailing F-16. One night in 2015, an American Viper pilot was running dangerously low on fuel over ISIS-held territory. The nearest KC-135, piloted by Captain Nathanial Beer from the 384th Aerial Refueling Squadron, responded immediately to a low-fuel distress call. "We were in the area of responsibility and were already mated with some A-10s," he recalled. "The lead F-16 came up first and then had a pressure disconnect after about 500 pounds of fuel," meaning that the F-16's fuel intake system had jammed. The Viper was no longer running on fumes, but 500 pounds was barely enough to stay airborne for the trip back home. Indeed, the starving F-16 would need at least *2,500 pounds* of fuel to return safely.

Given the nature of the fuel system jam, Beer's crewmen could only refuel the starving Viper in quick bursts, giving it enough fuel for approximately fifteen minutes of flight time. Thus, the KC-135 fueled the tethered Viper every fifteen minutes while escorting it away from enemy lines. The tandem F-16/KC-135 crept its way back to friendly airspace, untethering as soon as the F-16 could make its initial descent into the Allied airfield.

Had Captain Beer and his KC-135 not been there, the F-16 pilot would have had to eject over enemy territory. The 384th Aerial Refueling Squadron commander, Lieutenant Colonel Eric Hallberg, later praised the crew's heroism: "Knowing the risks to their own safety, they put the life of the F-16 pilot first and made what could've been an international tragedy into a feel-good news story." He added that what drove these KC-135 crewmen was: "A higher calling to be the best at the mission and take care of their fellow Soldiers, Sailors, and Airmen."

By 2015 the coalition's strategy evolved. In March, Kurdish and Iraqi forces began pushing back against ISIS, and airstrikes increased over Mosul. Yet the situation in Syria remained tenuous. The largest in-theater change came that summer when Turkey agreed to let US warplanes use Incirlik Air Base. In early August, the 31st Fighter Wing swung six F-16CGs (*Block 50*) from Aviano Air Base into Turkey. On August 12, 2015, those same Vipers flew the first US strike fighter sortie from Turkey against ISIS targets in Syria. By opening the runways at Incirlik, round-trip flight times into northern Syria decreased, while the Falcons' time on station increased exponentially.

As the deployment from Aviano unfolded, F-16s continued suppressing ISIS positions, supply lines, and oil wells. By late 2015 the campaign was moving into its "degrade" phase, focusing on ISIS's fast-expanding empire. The 31st Fighter Wing pilots, alongside F-15E Strike Eagles and international jets, hit moving convoys and command posts. That summer, for example, American F-16s bombed a Kurdish village where ISIS was assembling suicide vehicles. Another strike destroyed two vehicle-borne IEDs that ISIS militants had prepared near Mosul.

Meanwhile, back in Iraq, the training of Iraqi pilots also accelerated. Still flying their US-supplied F-16s, Iraqi Air Force pilots began flying missions over ISIS terrain with coalition support. Also in 2015, the US began rotating units from the Air National Guard. Thus, the 352nd Fighter

Wing from Fort Worth, flying their F-16C/Ds, took up alert missions in Turkey and Jordan. Across the region, F-16s flew thousands of sorties that year.

By 2016, Inherent Resolve shifted from "degradation" to "Counterattack & Defeat." On the ground, Iraqi forces prepared to recapture Mosul. In Syria, Kurdish allies closed in on Raqqa, the ISIS capital in Syria. American F-16s focused on supporting these final assaults, flying precision strikes in urban environments.

Along the Iraqi front, the Battle of Mosul began in earnest on October 17, 2016. F-16s from Balad Air Base (north of Baghdad) and support bases in Kuwait, Jordan, and Turkey were heavily involved. Falcons struck ISIS strongpoints, supply caches, vehicle suicide bomb factories, and hardened hideouts. They often flew at night to avoid Syrian-regime or civilian traffic. Iraqi leaders credited coalition air power for crippling ISIS defenses. One strike was confirmed to have "destroyed an ISIS fuel point near Huwayjah," while other strikes that day "destroyed 21 fighting positions" and vehicles near Mosul.

These strikes were coordinated with tight controls to protect civilians. Within the Viper's cockpit, pilots relied on satellite data relays and ground spotters. A junior F-16 weapons officer once described the focus on minimizing collateral damage: "We carried precision GPS-guided bombs and cued them on Iraqi city block coordinates handed down by ground controllers…every target ID took extra care." But the results were palpable. As General Bob Brown (OIR Air Component commander) later noted, the coalition's air campaign "helped Iraqi and Kurdish forces approach and enter Mosul more safely by softening ISIS positions."

Simultaneously, across the border in Syria, Kurdish-led forces laid siege to Raqqa. F-16s flew support missions, often alongside A-10s and the growing variety of UAV drones. They strafed enemy insurgents firing from buildings and struck mortar teams before they shot off their tubes. By mid-2017, Raqqa had fallen. In fact, throughout 2017, the coalition began launching more frequent and deliberate strikes as the caliphate's core infrastructure shrank. The year 2017 also saw the rise in high-tempo joint missions. F-16Cs often escorted B-52s and F/A-18s into combat. They also provided tactical air support for Special Forces troops and flew ground attack missions over northern Iraq and the Euphrates. They interlaced with Allied jets from the UK, France, and the Gulf states, under a tightly-regulated framework. Each Viper sortie was part of an evolving mosaic of aircraft types. Ultimately, it was the relentless pressure from jets like the F-16 that helped pin ISIS into its last-stand. Iraqi forces declared Mosul fully-liberated by July 2017, while the Kurdish-led Syrian Democratic Forces declared Raqqa freed by October 2017.

The final phase of Inherent Resolve—fittingly named "Defeat"—stretched across 2018 and into the years beyond. By then, ISIS no longer ruled over cities or provinces; instead, it was clinging to scattered redoubts in the Syrian wilderness and the rugged borderlands of Iraq. The F-16s, already veterans of the campaign's heaviest battles, adapted to the twilight of war. They launched not only from established hubs in the Gulf, but also from austere forward bases carved out of the desert. At Qayyarah West, once under ISIS's black banner, Vipers roared skyward on daily hunts for fugitives of the

Caliphate. The tempo had shifted: Gone were the massed frontlines of Mosul and Raqqa. Now, they had been replaced by fleeting targets—convoys slipping across the desert, caches hidden in villages, and depleted cadres plotting their next ambush. But the pilots understood that while the caliphate's territory had been broken, its ideology had not. The mission was no longer to smash a jihadist army, but to stalk an insurgency that thrived in the shadows.

But even as ISIS militants were on the run, other dangers persisted throughout 2018. Environmental conditions in the Middle East, for example, were known for their mercurial brutality and unforgiving relentlessness. In the summer of 2018, a violent sandstorm erupted near Ahmad al-Jaber Air Base in Kuwait. Straight-line winds of 91 miles per hour uprooted steel sunshades over the flightline—the first such maintenance disaster since Desert Storm. Seven F-16s from the 148th Fighter Wing (Minnesota Air National Guard) were crushed under collapsing hangars. Five of the seven Falcons were written off entirely. Normally, such a catastrophic loss would have grounded the squadron's mission, but the F-16 mechanics responded with ingenuity, purpose, and a fierce sense of urgency. Within days, engineers removed the tangled frames from the stricken F-16s while airframe experts cut away the mangled tails.

The 148th's maintenance chief, Chief Master Sergeant Ryan Gigliotti, recalled: "We quickly prioritized our efforts to get our fully mission-capable jets ready [and] fix the least-damaged jets first." Senior Airman Wyatt Struppler, on his first deployment as a crew chief, described learning every part of the Viper to fix the storms' damage. The ultimate triumph came when two hulks that were initially declared "beyond repair" were painstakingly rebuilt and returned to the sortie queue. Meanwhile, Lieutenant Colonel Matt Russell, an F-16 test pilot who flew the repaired jets, marveled: "It's pretty remarkable to get two jets flying that could have been shipped home in a box. They were able to add sorties to their Air Tasking Order in combat." Major William Carr, who had supervised the 148th's maintenance team during their deployment, added: "The pride felt by many Airmen when the two severely battle-damaged jets returned to combat operations was inspiring."

In total, the 148th flew 682 combat sorties during that 2018 deployment, totaling 3,899 flight-hours—even with depleted forces. Their story underscores the rarely-seen resilience needed to sustain a relentless air campaign against an asymmetric enemy.

By the end of 2019, ISIS had lost all its major territory. Iraq declared victory in December 2017 and the caliphate's Syrian enclave fell by March 2019. However, the F-16's saga did not simply end in 2019; it instead evolved into a long campaign of vigilance. From 2019 onward, OIR transitioned into an advisory mission, and many coalition jets (including F-16s) returned home or shifted to training roles.

Still, during these recent years, coalition air power has not been idle. F-16s continue to play a policing role. In January 2025, for example, the US Air Force reported American F-16s flying airstrikes over Iraq's Hamrin Mountains and against Iranian-backed militias—a testament to how the Falcon's combat history in OIR has extended into "mop-up" and deterrence missions. Throughout the spring of 2025, CENTCOM leadership emphasized the "global effort" to ensure ISIS's enduring defeat. As CENTCOM Commander, General Michael Kurilla, noted: "Partnered operations like these are critical to maintaining pressure on ISIS...the enduring defeat of ISIS relies on our coalition, allies, and partners." The Pentagon's Deputy Press Secretary, Sabrina Singh, echoed that America's residual forces in Iraq and Syria exist so that ISIS "can never reconstitute or resurge."

Beyond the headlines, however, OIR left an indelible impression on those who flew the F-16 into combat. Veteran commanders reflect on the intensity and unique nature of the mission. CENTAF chief, Lieutenant General Charles Brown, observed: "There is no doubt coalition air power has, and continues to, dramatically degrade [ISIS's] ability to fight and conduct operations." F-16 squadron leaders praised their pilots for precision effects that were unparalleled to anything seen from the days of Enduring Freedom or even Iraqi Freedom. As one weapons officer noted: "We're conducting the most precise air campaign in history...We're able to attrit [ISIS] and its capabilities anytime, anywhere."

Throughout the story arc of Inherent Resolve, there was little question that the F-16 lived up to its legacy. Its combination of range, precision, and adaptability proved indispensable. In the words of one crew chief, the F-16 after months of war "still needs a break now and then...they still need to be refreshed," yet they always answered the call. At this writing, OIR is slated to end by the Fall of 2025, but the US continues to monitor ISIS threats regionally. Nevertheless, the American F-16 fleet—rapidly-aging by modern aircraft standards—has played a pivotal role in the fight against the Islamic State. For more than a decade in the skies over the Levant, the

Viper flew thousands of sorties, dropped thousands of bombs, and demonstrated the Air Force's doctrine in action: Speed, flexibility, and overwhelming firepower. From their cockpits at 10,000 feet, F-16 pilots held "an invincible monopoly of asymmetric aerial firepower," as one analyst put it, enabling Iraqi and Syrian forces to reclaim their lands from the savagery of jihadism.

The tale of OIR's F-16s—from the shock of near-misses and mechanical failures to the jubilant homecomings—have become the stuff of legend in the Air Force community. Captain Nathanial Beer and his KC-135 boom-operator crew might not be F-16 pilots, but they savored the thrill of saving a flier who was seconds away from ejecting. The Minnesota Guardsmen still tell of the spring winds of 2018, and how they found purpose in piecing their Falcons back together under canvas shelters with little more than grit, improvisation, and elbow grease. They speak with pride of the rebuilt F-16s that "should have been shipped home in a box," yet flew 18 additional combat missions. This blend of triumph and sacrifice made the F-16 more than just a fighter jet. It had become a symbol of grit and resolve.

Arab States

For the Kingdom of Jordan, Operation Inherent Resolve was not a distant conflict but a war fought at the nation's very doorstep. When Jordanian F-16s joined the opening wave of coalition airstrikes in September 2014, their presence became a potent symbol of Arab resolve against the Islamic State. But within months, the campaign would exact a heavy toll that seared itself into Jordan's national consciousness.

On December 24, 2014, a Jordanian F-16AM from the

1st Fighter Squadron launched from Muwaffaq Salti Air Base for a mission over northern Syria. The jet, part of a coalition package striking ISIS positions near Raqqa, was flown by Lieutenant Muath al-Kasasbeh, a 26-year-old Jordanian Air Force pilot. During the sortie, his aircraft suffered a malfunction and went down near Raqqa. Kasasbeh ejected but was quickly captured by ISIS militants.

His downing marked the first coalition jet lost over ISIS territory. Within days, ISIS released images of the young pilot in captivity, and in February 2015 they unveiled a video showing his horrific execution by burning. The spectacle ignited outrage across the world, but in Jordan it unleashed a wave of fury and grief. King Abdullah II, himself a trained combat pilot, vowed that al-Kasasbeh's death would not weaken Jordan's resolve.

The response was immediate and ferocious.

Jordanian F-16s surged into the skies in a campaign of retribution locals dubbed "Mission Muath." Over three days, Jordanian jets flew dozens of sorties against ISIS targets in Syria, hitting training camps, weapons depots, and command posts. Coalition officials confirmed Jordan more than doubled its usual sortie rate in the days following al-Kasasbeh's execution. In an iconic image broadcast on Jordanian state television, F-16s returned with cockpits draped in the national flag, symbolizing both vengeance and unity.

Far from cowing the kingdom, the murder of al-Kasasbeh galvanized Jordan's role during Inherent Resolve. Amman intensified its involvement, with Jordanian pilots continuing combat missions from Muwaffaq Salti, often alongside US and Gulf allies. Al-Kasasbeh became both martyr and rallying cry. Streets and schools were renamed in his honor, his face painted on murals in Amman and Karak. Every sortie thereafter carried the memory of a young aviator who had become a national hero.

While Jordan fought ISIS on its northern flank, Egypt confronted the Islamic State in a different theater. In February 2015, ISIS's Libyan affiliate released a video showing the mass execution of twenty-one Egyptian Coptic Christians on a beach near Sirte. Within hours, the Egyptian Air Force launched retaliatory airstrikes against ISIS positions in Libya. Flying their American-built F-16C/Ds, Egyptian pilots struck training camps and weapons depots near Derna, demonstrating Cairo's willingness to project force beyond its borders in defense of its citizens.

Though Egypt's main fight was against insurgents in the Sinai, these Libyan airstrikes highlighted the regional reach of its F-16 fleet. In the campaign against ISIS, Cairo stressed sovereignty and unilateral action, underscoring that the Arab fight against Islamic extremism did not rely solely on Western leadership.

The Emirati Air Force, with its advanced fleet of *Block 60* "Desert Falcon" F-16s, played one of the most sustained Arab roles in Inherent Resolve. Operating from Al Dhafra Air Base, these Emirati F-16s joined US and coalition partners in striking ISIS convoys, infrastructure, and leadership nodes in both Iraq and Syria.

After al-Kasasbeh's capture, the UAE briefly suspended its participation, citing concerns about search-and-rescue capabilities. But within weeks, the Abu Dhabi government resumed operations, bolstered by US deployment of V-22

Ospreys for combat rescue. The UAE's commitment was more than symbolic: Its F-16s routinely flew alongside US aircraft in high-threat environments, showcasing the Desert Falcon's cutting-edge avionics and precision-guided firepower. The Emiratis also hosted Moroccan and Bahraini detachments, making Al Dhafra a hub of Arabian air power. Indeed, for the UAE, participating in OIR reinforced its status as a regional air power and a trusted partner in the Allied cause.

The Royal Bahraini Air Force, though modest in size, contributed its F-16s to coalition airstrikes from the outset. Flying alongside Gulf allies, Bahraini Falcons flew strike missions over Iraq and Syria beginning in late 2014. For the Bahraini pilots, their contribution carried political weight disproportionate to their size.

Indeed, Bahrain's involvement was less about sortie numbers than it was about the symbolism of Arab unity in the skies. Every Falcon painted in Bahraini colors that flew over ISIS territory reminded observers that the campaign was not only Western-led but regionally endorsed.

Meanwhile, at the farthest edge of North Africa, Morocco deployed a detachment of F-16C/Ds to the UAE in late 2014, committing their airmen to the fight against ISIS. Moroccan pilots flew combat missions over Iraq and Syria, striking ISIS targets with relentless fury. Their involvement was highly publicized in the Moroccan press, where the monarchy used their participation to underscore the kingdom's role as a moderate Muslim power standing against extremism.

In May 2015, however, tragedy struck.
A Moroccan F-16 was lost during a mission over Yemen

as part of a separate coalition operation against Houthi rebels. While not directly tied to Inherent Resolve, the loss underscored the risks faced by Moroccan aviators during a period of overlapping commitments. Still, Moroccan Falcons continued their role in the anti-ISIS campaign through 2015, often flying from Al Dhafra alongside their Emirati hosts.

From the Levant to North Africa, Arab F-16s became both weapons of war and instruments of legitimacy in the fight against ISIS. Jordan's agony over Muath al-Kasasbeh revealed the human cost of coalition warfare, while its ferocious response showed that the Jordanian spirit would not be broken. Egyptian airstrikes in Libya; the UAE's Desert Falcon operations; Bahrain's solidarity flights; and Morocco's expeditionary deployment demonstrated that the Arab world would not leave the campaign solely to Western powers.

In the skies above ISIS-occupied territory, the F-16's silhouette carried many flags, namely: Jordanian, Emirati, Bahraini, Moroccan, and Egyptian. Each sortie flew not only against ISIS, but also for the message that even majority-Muslim nations were willing to fight and bleed to extinguish the black flag of the radical caliphate.

Turkey

From late 2013 onward, Turkish F-16s kept a constant vigil over their 560-mile border with Syria. For many, the Turkish-Syrian border wasn't just a line on a map; it was a scar of history. The Ottoman collapse, Cold War maneuvering, and decades of hardened mistrust had turned neighbors into geopolitical adversaries. Throughout the latter half of the 20th century, Damascus openly sheltered the Kurdistan Workers Party (PKK), whose insurgents were fighting a guerrilla war inside Turkey. By 1998, the confrontation nearly erupted into a wider conflict. Turkey massed its troops along the Syrian border and issued an ultimatum: Expel the PKK leader Abdullah Ocalan or face the onslaught of Turkish firepower.

The resulting Adana Agreement, however, seemed to defuse the situation.

Under the terms of this 1998 agreement, Damascus renounced its sponsorship of the PKK, expelled Ocalan, and pledged to halt its support for Kurdish militancy. In return, Ankara pulled its forces back from the border. For more than a decade thereafter, the two governments cultivated a spirit of cooperation, trade, and even the beginnings of mutual trust—a respite that few would have predicted in earlier times.

But the eruption of the Syrian Civil War in 2011 quickly ended that period of détente. The relationship collapsed into open hostility, sharpened further in June 2012 when a Syrian SAM shot down a Turkish RF-4 Phantom over the Mediterranean. To Ankara, the shootdown wasn't merely an attack on an aircraft, but a violation of sovereignty and dignity. The response was swift and unequivocal: Turkey's Ministry of National Defense tightened their ROEs. From that moment on, any Syrian aircraft—fighter, helicopter, or drone—that strayed near the border would be declared a hostile contact.

Thus, the skies over Hatay and Gaziantep were no longer a neutral zone; they were a battlefield in waiting.

In this new climate of high alert, Turkish fighters patrolled the skies daily. On September 16, 2013, two F-16s intercepted a Syrian Mi-17 transport helicopter that strayed west of Hatay. After repeated radio calls went unanswered, one of the F-16s fired its AIM-9 Sidewinder and brought the chopper down over Syrian territory. The Assad regime admitted that its Mi-17 had accidentally crossed the border, but Ankara's point was clear: Any future incursions would be met with deadly force.

In the spring of 2014, the Turkish Air Force reiterated this policy with the warhead of an AIM-120 missile. On March 23, two Turkish F-16s from Diyarbakır scrambled to intercept a pair of Syrian MiG-23s approaching the Hatay frontier. According to Turkish accounts, one MiG heeded the warning and vectored back into Syrian territory; the other MiG, however, remained in Turkish airspace. The leading F-16 fired its AIM-120, striking the intrusive MiG which came crashing down on the Syrian side of the border. Turkish Prime Minister Recep Erdogan lauded the shootdown as a stern lesson: "If you violate our border, our slap will be hard," he told the nation.

As it turned out, the next "hard slap" from a Turkish F-16 would be directed at a low-intensity threat. On May 16, 2015, a pair of Ottoman F-16s engaged an unmanned aerial drone that had violated Turkish airspace over Hatay. Initially, Turkish authorities reported the intruder to be a Syrian helicopter, but Damascus and state media insisted it was only an unmanned surveillance drone. It later emerged that the drone was an Iranian-built Mohajer-4 UAV operated by Assad loyalists. The Mohajer-4 penetrated some seven miles into Turkish airspace, loitering for about five minutes before the F-16s intercepted it. Each F-16 fired an AIM-9 Sidewinder, one of which hit the drone, sending it into a downward spiral before it crashed onto the Syrian side of the border. This incident—the first known shootdown of an Iranian-built UAV by Turkish jets—reinforced the severity of Turkey's defensive posture.

By mid-2015, however, the Ottoman Vipers had transitioned from homeland air defense to expeditionary airstrikes against the Islamic State. A key turning point was the July 20, 2015 suicide bombing at Suruc, which killed nearly three dozen Turkish civilians. In response, Ankara joined the US-led bombing campaign against ISIS. On August 28-29, Turkish F-16s flew their first combat sorties as coalition partners, hitting ISIS-held targets in northern Syria. Official statements reported that a pair of Turkish warplanes struck four Islamic State targets north of Aleppo. These strikes—coordinated from bases like Incirlik—signaled Turkey's new willingness to prosecute ISIS militarily while still conducting their aggressive border-defense missions.

By late 2015, Turkey's air war along the border had begun colliding with Russia's intervention in Syria. Beginning in early October, Turkish F-16s would tally a number of run-ins with the Russian Air Force. On October 3-4, a flight of Russian Su-30s and Su-24s twice slipped into Turkish airspace near Hatay. Turkish F-16s on patrol responded by closing on the intruders; but the Russian Sukhois quickly turned away under warning. Over the next few days, Turkish air defenders remained on high alert. On October 6, Turkish F-16s intercepted a pair of Russian Su-30s that ventured close to the border. Reports indicate that at least one Su-30 briefly entered Turkish airspace and was tracked by two intercepting F-16s on patrol. One Turkish analyst noted that the Russian Sukhoi even locked its radar onto the Turkish jets for more than five minutes.

Moscow, for its part, claimed the border incursions were just "navigational errors" due to poor weather, but NATO's Secretary-General countered that such repeated incursions could not have been accidents.

Tensions escalated further in mid-October. For on October 16, a Turkish F-16 shot down another intruding UAV over Hatay. Witnesses described it as a small prop-driven aircraft, which US and Turkish officials later identified as a Russian-built Orlan-10 reconnaissance drone. Western media noted the wreckage matched Russia's Orlan series, and Turkish sources quietly pointed to Russia as the owner. The Kremlin, however, denied losing any aircraft that day, stating that "all its jets and drones had returned safely." In other words, Turkey shot down what was clearly a Russian-built spy craft; but Russia insisted it didn't belong to them. Still, the shadowy encounter underscored how fraught the situation was becoming as both sides danced around the blame.

The climax came on November 24, when two Turkish F-16s downed a Russian Su-24 fighter-bomber. Turkey's account was dramatic: The Su-24 had overflown Syrian rebel-held areas before inexplicably turning toward the Turkish border. Turkish command said they issued "ten warnings within five minutes" before the bomber allegedly crossed a sliver of Hatay airspace. When the incursion continued, one Turkish F-16 fired an AIM-9 Sidewinder, destroying the Russian jet from standoff ranges. Video footage immediately circulated showing the Su-24 plummeting in flames and its two pilots ejecting.

Russia's version of the engagement, however, was starkly different.

Within hours, Moscow announced that its fighter-bomber had "never violated Turkish airspace," remaining over Syrian territory the entire time. They even suggested that a rogue SAM or rebel ground fire, not a Turkish jet, might have brought the plane down. Erdogan quickly labeled these claims as "lies" and released radar images showing the Su-24 briefly over Hatay. The White House later confirmed that the plane had indeed clipped Turkish airspace "for a matter of seconds"—about 17 seconds according to Turkey's letter to the UN—during which it was shot down.

One of the Russian crewmen was killed by ground fire after ejecting; the other evaded capture. A desperate search-and-rescue mission ensued, during which a Russian Marine was killed when his helicopter came under rebel fire. Moscow erupted in a fury: Vladimir Putin denounced the incident as a "stab in the back" and blamed Turkey for abetting terrorists. The Turkish government, however, defended itself by claiming they had simply upheld their sovereign airspace. For months afterwards, the two countries severed ties and imposed mutual sanctions.

Through 2013–15, Turkish F-16 squadrons fought an unusual air war. Unlike other coalition partners who mostly bombed ISIS targets, Turkey's F-16s were constantly poised to engage enemy aircraft. They downed Syrian helicopters and fighters, destroyed drones, and eventually locked missiles on Russian warplanes—all in the name of defending Turkey's borders. In exchange, Turkey suffered a severe diplomatic fallout with Bashar al-Assad's patrons. But from the Turkish perspective, every shootdown over Hatay underscored a simple truth: Every hostile shadow in the Turkish sky would be met with deadly fire.

Following the Su-24 shootdown of November 24, 2015, the skies over the Syrian-Turk border were relatively quiet for the next five years. But as the border patrols dragged on into 2020, the Ottoman F-16s suddenly found themselves at the epicenter of another high-stakes aerial conflict.

In early 2020, the simmering conflict in Syria's Idlib Governate erupted into an open confrontation, marking a sharp turn from Turkey's cautious engagement to direct air-superiority operations.

On February 27, 2020, a joint Syrian-Russian airstrike obliterated a Turkish Army convoy in Balyun, killing some 34 Turkish soldiers. It was the deadliest blow against the Turkish military since entering Syria, and it shattered any remaining restraint on retaliation. Within days, Turkey launched Operation *Spring Shield*, signaling a full-throttle return to aerial warfare. The skies over Idlib became a battlefield once more, this time with Turkish F-16s answering fire with fire.

On March 1, 2020, two Syrian Su-24s faced off against Turkish air defenses—and fell. Turkish F-16s shot down both aircraft over the Idlib province, a result confirmed by both sides. Syrian state media confirmed the loss and that their pilots ejected safely. Turkey declared the Su-24s had threatened its airborne assets and medical outposts near the frontier. The names of the Turkish pilots have not been publicly disclosed; but official statements from the Ministry of Defense framed the action as a protective measure, essential to defending Turkish lives and strategic interests. In diplomatic corridors, NATO noted the engagement as a rare escalation—Turkey's F-16s had reignited the role of air superiority that they had typically avoided in recent years.

Two days later, the air clashes continued. On March 3,

a Turkish F-16 engaged and destroyed a Syrian L-39 light attack aircraft over Idlib. Local monitors reported the two-man crew was killed, but Syrian authorities gave conflicting accounts: One said the downed crew had survived; others indicated that at least one of them perished.

These three shootdowns—two Su-24s and one L-39 across 72 hours—were more than tactical victories. They marked a strategic shift: Turkey's F-16s were flying not just to defend their homeland airspace, but to dominate the skies above Idlib. In the process, they achieved air superiority over one of Syria's most disputed warzones.

While external commentaries from NATO and Western analysts condemned Turkey's swift escalation, their operations underscored the F-16's lethal edge. No pilot testimonies have been published regarding these 2020 aerial victories, preserving the highly-discrete nature of the operation. Still, the Turkish government's line was clear: "We will not be silent," echoed in the missile blasts that downed each Syrian bandit.

This campaign of March 2020 marked the most intense aerial engagement between Turkish F-16s and Syrian jets in five years. Yet it re-established Turkey's willingness to assert its airspace and strategic buffer zones by lethal means. The escalatory tone of shooting down multiple Syrian aircraft within three days sent a blunt message to Assad and his backers.

The tactical consequences of the March shootdowns were immediate. Syrian fixed-wing activity over the Idlib battlespace dropped off precipitously, and Assad loyalists began relying more on artillery and Iranian-backed auxiliaries to press ground attacks—avenues of power less-exposed to Turkish air interdiction. Turkish commanders, for their part, doubled down on the tactics that had yielded results: Drone-heavy strikes backed by standoff fighter patrols and aggressive counter-air operations. By the time the March 5 ceasefire took hold, the Syrian offensive in the region had stalled.

But the years after 2020 did not bring a clean peace. Along the frontier, Turkish F-16s continued their quick-reaction alerts, flying CAPs over Hatay and Gaziantep provinces, and responding to periodic "hot" scrambles when Syrian or Russian aircraft tested the borders. The choreography of modern limited war—radar locks, hotline calls, and patrol zones edged up against political red lines—became routine. But the character of Turkish air action remained consistent with the *Spring Shield* template: Impose local air denial on the SyAAF when they threatened Turkish forces or proxy units, and use standoff tools to keep the ground fight at arm's length. Whenever flare-ups occurred (such as artillery strikes, drone losses, and the occasional air-to-air exchange) the F-16s were there as both a sword *and* shield, a deterrent against Syrian intrusion.

Seen against the greater arc of the Syrian conflict, the Turkish F-16s' air-to-air engagements in March 2020 were an aberration…and a lesson. Most coalition air power expended over Syria and Iraq since 2014 had targeted ISIS: An enemy without a modern air force, dangerous on the ground, but nearly absent in the air. Turkey's battles over Idlib were different: A state-on-state duel nested within a greater proxy war, under a sky frequently patrolled by Russian jets. In that environment, every radar contact was political, every missile shot was a message to at least three audiences: Syria, Russia, and the opposition forces. But the Vipers' air-to-air victories did more than kill enemy planes; they created negotiating space that Turkey used in Moscow a few days later. The ceasefire that followed on March 5, 2020 was fragile and incomplete, but it validated the results that Turkish air power had achieved in the skies over Asia Minor.

As with every chapter of the Syrian Civil War, alternative framings persist. Damascus insisted its aircraft were on counter-terror missions in sovereign airspace and portrays the 2020 losses as the consequence of Turkish aggression on behalf of "terrorist" proxies. Turkey answered that its jets were defending Turkish assets from hostile activity, and that the strikes were lawful self-defense. Russia, while denying responsibility for any specific air-to-air engagements, made clear that it viewed Turkey's fixed-wing operations over Idlib as destabilizing. The public records (including ministry statements, wire-service reports, and dueling communiqués) do not resolve every contradiction. But all reputable sources agree on the central facts: Two Syrian Su-24s were shot down by Turkish F-16s on March 1; one Syrian L-39 was shot down by a Turkish F-16 on March 3; and a Russian-Turkish brokered ceasefire took effect on March 5, shifting the fight from the sky to the negotiating table.

In the end, the *Spring Shield* vignette is a reminder that the Syrian air war chapters are multifaceted: There were long, grinding counter-ISIS sorties flown by many nations; there were Russian campaigns of punishment from on high; and there was, briefly and decisively, a state-on-state confrontation over Idlib in which Turkish F-16s restored deterrence by shooting down three Syrian jets over the course of three days.

Belgium and the Netherlands

When the coalition called for pilots in 2014, it wasn't the great air arms that answered first. Rather, it was the smaller, highly-professional air forces of Western Europe—particularly Belgium and the Netherlands. With a handful of Vipers and a few hundred airmen, they stitched themselves into an enormous American-led air campaign, bringing disciplined ROEs and a blunt message that Europe would not stand by idly while the Islamic State brutalized the Middle East.

The Royal Netherlands Air Force (RNLAF) moved with deliberate intent. In September 2014, the Dutch government approved sending their F-16s to join the campaign in Iraq, purportedly on the basis of Iraq's request for assistance. Six Dutch F-16s were tapped for deployment. Though their numbers were small, their operational tempo was high. The RNLAF detachment flew from bases in Jordan under CENTCOM control and quickly settled into a pattern of CAS and low-level airstrikes against ISIS targets. By late 2015, Dutch Vipers had flown hundreds of sorties and dropped hundreds more bombs. Following the operational success of the RNAF in combat, The Netherlands extended its mission into 2016, striking ISIS targets in both Syria and Iraq.

From an operational standpoint, the Dutch approach was a textbook example of integrative coalition air power: Plug into the Joint Forces Air Component Command, internalize the coalition's targeting cycle, and execute with clinical efficiency. Dutch pilots flew day and night, often in pairs, using laser-guided and GPS-guided weapons to limit collateral damage. But even careful techniques and disciplined airmanship couldn't mitigate the tragedies of war. Investigations soon revealed that Dutch airstrikes had inadvertently caused civilian casualties. For example, a miscalculated airstrike near Hawija in April 2015 drew fierce condemnation, prompting both a public apology and subsequent reforms within the Ministry of Defense. The episode reminded Dutch pilots and politicians that precision firepower, while necessary, could not always prevent collateral damage.

Belgium's contribution, meanwhile, was likewise modest in quantity but tremendous in its impact. Brussels rotated small detachments of F-16s (typically four to six jets) into theater under the auspices of Operation *Desert Falcon*. Belgian Vipers flew from Jordan and, at times, from air bases nearer the action, performing CAS, reconnaissance, and interdiction missions. During those years of deployment, Belgium's F-16s punched well above their weight class. Indeed, across every rotation, they recorded several hundred sorties, yielded thousands of intelligence products, and released a modest number of precision-guided weapons when tasked. A 2021 after-action summary reported that a recent Belgian detachment had flown 302 missions. These sorties totaled more than 3,000 flying hours, with 29 precision-guided bombs delivered and 1,613 intel products generated—numbers that highlighted the detachment's high-tempo, intelligence-driven battle rhythm.

Belgians brought to the fight a particular strength: The integration of their reconnaissance and targeting flow into coalition systems. Belgian crews regularly executed dynamic targeting missions and produced data that fed into American-led strike packages. For a smaller air force like Belgium's, the learning curve was steep and the operational tempo was punishing, but their air crews responded with a strong and steady professionalism.

Public accounts of Belgian deployment rotations emphasize their endurance—long nights in the cockpit, long mission planning cycles, and the steady stream of surveillance and reconnaissance that made later airstrikes more discriminating.

Together the Dutch and Belgian F-16s embodied a certain European model of air engagement: Low-density deployments with highly trained crews, strict political caveats, and a heavy emphasis on precision intelligence. Their sorties tended to be smaller and deliberate rather than massed. Typical armaments included GBU-12/16 laser-guided bombs or JDAMs tied to coalition targeting. Meanwhile, reconnaissance pods, targeting officers, and coalition AWACS helped tighten the kill chain. This approach fit the restrictive parameters imposed by the home country, but it also leveraged the unique strengths of modern multirole fighters: Speed to target, the ability to shift from reconnaissance to strike missions during a single sortie, and the capacity to integrate quickly into a multi-national strike package.

Yet both the Belgian Air Force and the RNAF had their institutional limits. Just as they had discovered during the Libyan No-Fly Zone missions, these smaller air forces could only surge their effects for a limited time. Belgium and Holland, like other smaller partners, repeatedly faced a problem that NATO planners had predicted: Stocks of precision-guided munitions were finite, and replenishing them during a multi-national war effort required political and logistical choreography. Consequently, the Benelux experience helped drive later reforms within the coalition to improve and streamline the delivery of shared munitions.

On the ground the story of the Benelux F-16s was as

much about people as it was about the jets. Dutch and Belgian pilots departed the comfortable confines of Western Europe to fly repeated short-duration combat sorties, often launching in the middle of the night, only to return and brief within hours before going into their mandatory rest cycles. Mechanics worked 24-hour shifts to keep the Vipers mission-ready; intelligence officers fused sensor feeds into targetable recommendations. For these small contingents, each sortie represented a collective effort of dozens of airmen—a human ledger rarely visible to the casual reader but indelible to those who served.

However, no military campaign is free from controversy. In The Hague, news of the civilian casualties at Hawija damaged public trust and forced a reckoning within the RNAF over targeting/vetting procedures and operational transparency. The Dutch Ministry of Defense acknowledged these shortcomings and adjusted procedures for mitigation and reporting collateral damage. In Brussels, meanwhile, every extension of the F-16s' deployment prompted another political debate: How many jets, for how long, and under what caveats? Those debates reflected the core tension of military operations controlled by democratic governments: Public scrutiny influences strategy as much as battlefield realities.

All told, the Belgian and Dutch F-16 deployments will be remembered for their restraint and effectiveness. They helped blunt ISIS maneuvers and fed an intelligence pipeline that made the coalition's airstrikes more effective. They also taught uncomfortable lessons about the limits of precision firepower in urban combat and the burden of sustainment on small air forces. The Low Countries did not lead the coalition by size, but they helped lead it in terms of professional craftsmanship, showing that trained pilots, disciplined ROEs, and a steady recon-to-strike rhythm can have disproportionate effects within a multinational campaign.

Yemeni Civil War

The Yemeni Civil War erupted in 2014, with the rise of the Houthi insurgency—an Iranian-backed militant group based in northern Yemen. On January 22, 2015, Yemeni President Abdrabbuh Mansur Hadi was forced to resign by the Houthi insurgents and was placed under house arrest. The Houthis named a Revolutionary Committee to assume the powers of the presidency, but Hadi soon escaped house arrest, fleeing to his hometown of Aden. Now beyond the reach of his captors, Hadi rescinded his resignation, denounced the Houthi takeover, and called upon Saudi Arabia for a military intervention.

On March 26, a Saudi-led coalition launched Operation *Decisive Storm*, an air campaign designed to blunt the Houthi advance and restore Hadi's government. The coalition brought together not only the Royal Saudi Air Force, but also the expeditionary strength of the United Arab Emirates, Bahrain, and the Kingdom of Jordan. For these Arab partners, the Yemeni campaign represented both an assertion of regional security and a test of their modernized F-16 fleets.

From the outset, the UAE Air Force assumed a leading role second only to Saudi Arabia. The Emiratis committed their most advanced fighters: The *Block 60* Desert Falcons and Mirage 2000s. Operating from bases in the UAE, southern Saudi Arabia, and forward airstrips in Yemen itself, Emirati squadrons delivered precision-guided strikes on Houthi positions, air defenses, and supply lines.

Emirati jets were crucial in the Battle of Aden, providing tactical air support to coalition and Yemeni government forces during the brutal fight to wrest control of the city from Houthi guerrillas. Emirati pilots flew suppression missions against artillery and coastal missile batteries, enabling friendly amphibious landings and resupply by sea. A later RAND analysis noted that UAE ground forces, supported by air power, became "the most effective coalition partner on the ground," holding captured territory and stabilizing southern strongholds.

Still, the war exacted a toll on the Emirati air fleet.

On March 14, 2016, a UAE fighter was lost during a combat mission, killing both crewmen on board. The official cause was listed as a "mechanical error," but it underscored the risks of sustained expeditionary operations. Abu Dhabi honored the fallen aviators as national heroes, adding their names to a widening ledger of Emirati sacrifices in the fight for regional security.

By 2019, the UAE gradually scaled down its combat role, declaring its "war in Yemen is over" in terms of major frontline operations. But even as their footprint contracted, the Desert Falcons had nevertheless made their presence known. They had flown hundreds of sorties, cementing their reputation as a regional air power capable of sustained, independent, expeditionary warfare.

For Bahrain, Yemen represented both solidarity with Gulf allies and a rare chance to deploy its F-16s in combat. The Royal Bahraini Air Force committed a detachment of F-16C/Ds to the coalition, flying strike missions against Houthi infrastructure and convoy targets.

The risks were evident from the start. On December 30, 2015, a Bahraini F-16 crashed while returning from a mission over Yemen, coming down inside Saudi territory.

Coalition officials quickly confirmed the loss, attributing it to a technical malfunction rather than hostile action.

Though Bahrain's detachment was modest compared to Saudi or Emirati numbers, its presence carried political weight. Every Bahraini sortie reinforced the Gulf Cooperation Council's unified stance. In practice, Bahraini pilots joined composite strike packages with Emirati and Saudi aircraft, hitting radar stations, ammunition depots, and frontline troop concentrations.

For Jordan, fresh from the trauma of Muath al-Kasasbeh's murder by ISIS in early 2015, participating in the Yemeni Civil War symbolized both resilience and solidarity. Amman pledged F-16s to the coalition and began flying missions from Saudi bases. Jordanian fighters struck supply lines, air defense positions, and assembly points, drawing on the Royal Jordanian Air Force's hard-won experience from recent operations against ISIS.

Though Jordan has never released a detailed sortie count, its participation was noted in coalition communiqués and Western wire reports. These airstrikes underscored Jordan's role as a reliable partner, willing to project airpower beyond its borders despite its own internal security burdens.

Arab F-16s were central to Operation *Decisive Storm* and its follow-on mission, Operation *Restoring Hope*. They flew interdiction sorties, battlefield strikes, and CAS missions to enable coalition ground offensives. But the combined efforts of these Emirati, Bahraini, and Jordanian airmen were more than just symbolic. They demonstrated that Gulf and Levantine air forces—who had previously relied on NATO and Eastern Bloc patronage—were now capable of conducting sustained expeditionary warfare on their own terms. The UAE's Desert Falcons in particular showed how a small state could project air power with world-class fighters and integrate them into amphibious and special operations.

Yet the Yemeni experience also highlighted the limits of air power absent a political solution. F-16s could destroy Houthi convoys, silence artillery, and shield ground allies, but they could not settle the war's deeper fractures: Collateral damage, sectarian divides, tribal rivalries, and external geopolitics.

In the end, Arab Vipers wrote their chapter in the skies over Yemen's civil war. They stood not only as instruments of resolve and symbols of regional coalition-building, but as witnesses to the human cost of a war where air supremacy couldn't deliver peace.

An Emirati F-16 "Desert Falcon" in Dubai, 2012. *Aeroprints*

Turkish F-16 in flight, 2014. The Turkish Air Force has made extensive use of its F-16s since the 1990s. Turkish Falcons have seen combat from the territorial islands of the Aegean Sea to the badlands of the Levant. *Jim Van de Burgt*

An F-16 from the Hellenic Air Force, 2024. These Greek F-16s have often intercepted Turkish aircraft over the Aegean Sea as part of the ongoing territorial water/island dispute between the two Mediterranean powers. *Jim Van de Burgt*

Chapter 8:
Brushfires

By the early twenty-first century, the F-16 had flown in nearly every major conflict involving Western air power. From the Bekaa Valley to Desert Storm, from the Balkans to the skies over Afghanistan, its record was etched into the modern history of air combat. Yet beyond these headline engagements stood a quieter, more fragmented story: The brushfire wars wherein the Viper still played a decisive role.

These were not the wars of sweeping coalitions or lightning campaigns. Rather, they were border skirmishes, regional rivalries, flashbang insurgencies, and internal upheavals. In these environments, the F-16 proved just as relevant as in the major conflicts. It patrolled the tense skies between Greece and Turkey, guarded Venezuela against violent coups and narco-terrorists, hunted enemy fighters along the Pakistani border, and even found a new lease on life in the Ukrainian Air Force.

Taken individually, these episodes seldom commanded the same gravitas as the air campaigns over Iraq, Kosovo, or Syria. But together, they underscore an essential truth about the Fighting Falcon: Its enduring adaptability and combat utility across any theater of war. The following case studies explore these overlooked chapters. They are lesser-known, perhaps, but no less vital in understanding how the F-16 has earned its legacy as the world's most versatile and battle-tested fighter.

Pakistan

In 1977, Pakistan was thrust into a new political era when General Muhammad Zia-ul-Haq toppled the country's civilian president in a swift coup d'état. Zia's rule was marked by the imposition of austere Islamic law and a determined drive to rebuild Pakistan in the wake of its defeat in the Indo-Pakistani War of 1971. Under Zia's leadership, military strength became a top priority.

Two years later, fate handed him an unexpected opportunity.

When Soviet forces invaded Afghanistan in December 1979, Washington saw a vital partner in the Islamic Republic of Pakistan. Zia's policies—including his anti-Communist views, his willingness to host Afghan refugees, and the reassertion of military power—dovetailed nicely with America's strategy of bleeding the Red Army dry in the Hindu Kush. Thus began one of the most fascinating and under-reported partnerships of the Cold War: US and Pakistani cooperation in supporting the Mujahideen, who based many of their operations from the refugee camps along Pakistan's western border.

In response, the Soviet and Afghan Air Forces began crossing into Pakistani airspace, raining fire on the cross-border refugee camps. The Pakistan Air Force (PAF) scrambled what planes they had—mostly Shenyang J-6 fighters (Chinese-built copies of the MiG-19). Though capable of supersonic flight, these aging interceptors were easily outmatched. Thus, for a time, the frontier skies over Pakistan belonged to the Soviet-Afghan strike fighters.

In 1981, however, the imbalance began to shift. Through persistent lobbying, Zia persuaded the US to sell Pakistan forty brand-new F-16As, along with a pair of two-seat F-16Bs. Deliveries trickled in between 1983 and 1986.

These early-block Vipers, however, lacked the radar-guided BVR missiles, forcing the pilots to fight close-quarters using the AIM-9 Sidewinder.

By 1986, the Pakistani Vipers of No. 9 and No. 14 Squadrons had been declared ready for combat. As the Mujahideen endured a Soviet offensive in the Panjshir Valley, the Pakistani F-16s took flight on their first combat patrols.

The balance of power along the frontier was about to change.

On May 17, 1986, radar operators at Badaber picked up two fast-moving contacts crossing into Pakistani airspace near Parachinar, their vectors angling toward the local refugee camps. GCI scrambled a pair of F-16s from No. 9 Squadron to intercept. Leading the patrol that day was the squadron commander, Abdul Hameed Qadri, with Mohammad Yousaf covering his wing.

The intruders were identified as a pair of Afghan Air Force (AAF) Su-22M3 strike fighters, their swing-wings swept for speed as they roared low over the rugged frontier. Under the J-6, these intruding aircraft would have easily escaped. But with the F-16, Qadri and Yousaf now had the speed and agility to run them down. Closing from behind and above, Qadri armed his AIM-9 Sidewinder, its seeker head growling as it locked onto the heat signature of the trailing Su-22.

At a range of less than two miles, Qadri pulled the trigger.

The Sidewinder erupted from his wing, smashing into the Sukhoi's afterburner with a brilliant burst of flame. The stricken fighter pitched up and began to atomize as it spiraled into the mountains. The second Sukhoi jinked hard right, diving fast for the border. Qadri gave chase, throttling his F-16 through the mountain air. Closing fast, he lined up for a gunshot, firing a multi-round burst from the autocannon. Tracer rounds painted a quick arc into the bandit's fuselage, followed by a heavy plume of smoke, marking the end of the second Su-22. As Qadri recalled: "I quickly rolled back and fired a three-second burst on the exiting Su-22. I stopped firing when a trail of smoke and flash from his aircraft confirmed a lethal kill."

It was Pakistan's first taste of modern air combat in the Soviet-Afghan war. Two enemy aircraft had been destroyed in the space of a few minutes, achieved with the precision of the F-16's speed and firepower.

The triumph echoed throughout the PAF.
For the first time, Soviet and Afghan pilots realized that their intrusions into Pakistani airspace would not go unanswered...and would be met with deadly force.

Still, these incursions continued throughout the spring of 1987. On the morning of March 30, for example, Pakistani radar stations lit up with an unusual contact: An Antonov An-26 transport, flying low towards the border. It remains unclear whether this An-26 was exclusively operated by the AAF, or whether Soviet pilots had been flying/advising it. What is clear, however, is that its presence was unwelcomed.

From PAF Base Sargodha, No. 9 Squadron scrambled Abdul Razzaq Anjum in his F-16, alongside his wingman, Sikandar Hayat. GCI vectored them northwest on a high-speed intercept. This An-26 was large and slow (presumably an easy target) but everything about it boded danger. At best, it was a transport; at worst, it was flying a reconnaissance mission for a forthcoming airstrike.

As Razzaq later described it: "The vector given by the

controllers started the flow of adrenaline. All the preparatory actions were over in less than 30 seconds. When I brought the target into the TD box at 3-4 [nautical miles], I realized that it was a slow moving, larger aircraft...the minimum range cue was lying close to 4,000 feet. Effectively, I had no more than a 1.5-second firing window available."

At that moment, Razzaq fired his AIM-9 Sidewinder.

Sikandar Hayat simultaneously fired his own AIM-9, and both missiles found their mark. The now-stricken An-26 spiraled downward into the mountainside near Miranshah, killing all 39 aboard.

Afghan media immediately protested, claiming the aircraft was a civilian transport carrying women and children to Khost. Pakistani authorities, however, insisted it was military...or at least was operating in a military capacity. Islamabad also maintained that the plane had violated Pakistani airspace and had ignored orders to land.

Still, the incident was monumental for the PAF. It was the first, large-scale "transport" kill achieved by Pakistani F-16s. For the pilots, however, the moment was more visceral. Razzaq's firing window closed quickly. Thus, the speed of decision (and the weight of its consequences) became amplified within mere seconds. And when the An-26 went down, it was a reminder that Pakistan's border skies would be fiercely defended for the remainder of the conflict.

The following month, the air war grew even more intense. On April 16, 1987, F-16s from No. 14 Squadron engaged yet another Su-22 near Dera Ismail Khan. The Squadron Leader, Badr-ul-Islam, chased down the fighter-bomber flown by Afghan Lieutenant Colonel Abdul Jameel. Gun camera flashes and tracers quickly lit up the sky. As expected, the Sukhoi went up in smoke; but Jameel ejected, and was promptly captured after landing on Pakistani soil. This marked Pakistan's third confirmed air-to-air kill of the war.

Days later, tragedy struck amidst the string of Pakistani victories. On April 29, a pair of F-16s from the No. 9 Squadron swept in to ambush four MiG-23s from the Soviet 120th Fighter Regiment, which had just bombarded the refugee camps in Djaware. As the Soviet flight leader, Lieutenant Colonel Pochitalkin, threw his formation into evasive maneuvers, he glanced downward to see a fireball spiraling down towards the earth. At first, he thought it was a fallen comrade...but this flaming wreckage was no MiG.

It was an F-16.

The stricken Viper belonged to Lieutenant Shahid Sikander Khan. In the split-second chaos of the intercept, his aircraft had been fatally struck by an AIM-9 Sidewinder, fired by his own wingman. The weapon, designed to seek out the hottest signature in the sky, had acquired Khan's jet instead of its intended target.

Khan ejected, drifting down onto the Afghan side of the border. For hours, his fate was unknown. But, fortunately for him, he was recovered by some local Mujahideen who spirited him (and the wreckage of his F-16) back into Pakistani territory.

Soviet sources later claimed that Khan had been downed by one of the escaping MiG-23s...but none of the MiGs had been carrying air-to-air missiles that day. For the PAF, however, the truth was much simpler: Friendly fire. The fog of war had claimed one of their own.

By the dawn of 1988, as Soviet ground forces began

withdrawing from Afghanistan, the air war grew fiercer. The Afghan regime was collapsing, and Moscow was trying to keep it afloat with an eleventh-hour series of "Hail Mary" bombardments. Indeed, for much of 1988, the skies over the western frontier were a nightly blaze of tracer rounds, missile fire, and blinding flares.

On the night of August 8, 1988, one of the conflict's most fateful duels unfolded. Colonel Alexander Rutskoy, a decorated Su-25 attack pilot, led a four-ship raid against the refugee camp at Miranshah. Slow but heavily armored, the Su-25 was built to absorb punishment. But that night, as Rutskoy rolled his formation towards their target, two F-16s from the No. 14 Squadron swept down from above.

Rutskoy wheeled the Su-25 into a desperate turn, hoping to draw the Pakistani fighters off and trusting that if his tailpipe faced away, any incoming missile would lose its lock. But the AIM-9 was no ordinary heat-seeker. It was an all-aspect killer, able to home onto any heat signature from any angle. When the missile streaked in, fired by Squadron Leader Athar Bokhari, its proximity warhead detonated with a surgical fury.

Rutskoy's "flying tank" broke apart mid-air.
The colonel ejected into the darkness, drifting down onto Pakistani soil, whereupon he was captured by local troops. For Islamabad, Rutskoy's capture was a politico-military triumph. For Moscow, it was an international humiliation.

Yet diplomacy prevailed.
In time, Rutskoy returned to the Soviet Union as part of a prisoner exchange. He would be designated a "Hero of the Soviet Union," rise to the rank of general, and (astonishingly) serve as Vice President of Russia under Boris Yeltsin, before leading an abortive coup attempt in 1993.

Thus, the brief encounter between a Pakistani F-16 and a Soviet Su-25 not only altered the balance of a single night's air battle, it touched the fate of a nation, and the fate of one man who would stand at ground zero of Russia's own tumultuous future.

In the weeks following Rutskoy's dramatic shootdown, the Soviets wanted revenge. Thus, in a show of strength, a flight of MiG-23s thundered across the border near the Kunar Valley on September 12, 1988. Most of them were carrying bombs designated for the refugee camps below; but a few were fitted with R-24 long-range missiles, hoping to bait any lingering F-16s into a dogfight.

At 32,000 feet, the trap was set.

Two F-16s from No. 14 Squadron took the challenge, flying in from 11,000 feet, but masked by the jagged terrain. While Soviet radars searched the skies above, the Vipers clung to the earth below, invisible amidst the clutter.

Finally, one of the Vipers broke high towards the offending MiGs. Lieutenant Khalid Mahmood fired his Sidewinder from a steep climb. Arcing through the sky, the angry AIM-9 tore through the trailing MiG-23. The wounded fighter stayed aloft briefly, only to crash in flames a moment later.

Startled by the sudden burst of missile fire, the Soviets broke formation. Two MiGs peeled off to engage the incoming Vipers. The ensuing clash was brief, afterburners roaring at full power as tracers lit up the mountain sky. The Pakistani F-16s claimed two more kills, yet Soviet records denied any losses—an enduring dispute in the fog of the border war.

On November 3, 1988, the skies over the Afghan

border yielded one final victory. Again, it was Khalid Mahmood at the controls. This time his quarry was an AAF Su-22M4K, skirting along the western frontier. A sharp maneuver, a perfect lock, and another Sidewinder struck true. The Sukhoi disintegrated, its wreckage a final testament to the PAF's vigilance.

With that shot, the ledger of confirmed victories closed. Pakistan formally credited their F-16s with ten aerial kills: four Su-22s, three transports, one Soviet Su-25, and two others contested in the record. Soviet archives, meanwhile, acknowledged only six losses: Four Su-22s, one Su-25, and a single An-26. But even by their measure, the frontier skies had been bloodied.

Today, rumors persist of other aerial victories, whispered among pilots and storytellers. Such tales feature intrusive bandits being chased back into Afghan airspace—downed, but never officially acknowledged for fear of widening the war.

But one fact is beyond dispute: Through two and a half years of relentless trial (1986-88), the PAF Vipers stood their ground, proving that Pakistan's skies would never again be violated without consequence.

Reflecting on the skirmishes from 1986-88, it's clear that the ad hoc air war has etched itself into the doctrine of the Pakistan Air Force. For the first time, their pilots had gone toe-to-toe against Soviet combat squadrons...and prevailed. The F-16, once derided as too costly and politically fragile, had proven itself a guardian of Pakistan's sovereignty, able to deter intrusions with precision firepower and ferocity.

Still, the lessons were hard and enduring. Pakistani planners internalized the value of forward-deployed CAP stations and honed the art of terrain-masking tactics against numerically-superior adversaries. The border skirmishes also emphasized the need for split-second decision-making in the cluttered, ambiguous battlespace of the Hindu Kush. Just as important, the "friendly fire" incident from April 1987 underscored the human limits of technology, reminding pilots that vigilance and discipline were as vital as their onboard instruments.

In the years that followed, the aura of these encounters elevated the F-16 to a near-mythical status in the PAF. It was no longer simply an American import; it was *their* Viper, baptized in the crucible of the Soviet-Afghan war. The aircraft's combat record became a touchstone for Pakistani air doctrine, influencing how squadrons trained, how air defense grids were layered, and how deterrence was communicated to neighboring adversaries.

Internationally, the episode cemented Pakistan's reputation as a regional air power. Few air forces could claim to have bested Soviet aircraft in open combat. Fewer still could claim to have done so with consistency. By 1989, the F-16 in Pakistani service gave proof that, in the hot skies over South Asia, the PAF was a force to be reckoned with.

Throughout the 1990s, Pakistan shifted its focus back to the eastern border with India. For most of the decade, the security situation along the Indo-Pakistani frontier had been relatively quiet, although tensions remained high. That tenuous peace, however, was abruptly shattered in 1999 with the onset of the Kargil War.

When the conflict erupted that spring, the PAF found itself tasked with a delicate mission: Deter a larger Indian air campaign while avoiding the exact conditions that would lead to such an escalation. The F-16 was now Pakistan's premier fighter, flown by pilots with years of experience flying frontier patrols. Despite these credentials, however, the Vipers would not be employed as the spearhead of offensive strikes, but as a shield of deterrence. From bases behind the Line of Control, Pakistani F-16s mounted cyclic fighter patrols, hoping to convince New Delhi that any broadening of the war would carry unacceptable risks.

These airborne sentinels maintained a watchful eye, vectored by radar nets and forward air controllers, even as both sides tried to keep the war limited to the ground. Scholarly accounts of air power in the Kargil War have emphasized this hedged approach: Air power was used sparingly because both governments feared escalation.

And the Pakistani F-16s were part of that careful calculus.

Nevertheless, logistics and political issues constrained the Vipers' deployment. PAF maintenance cycles had been suffering for years under various logistical and procurement issues of the 1990s. Thus, when the Kargil War began, sortie generation rates suffered. And, at times, the F-16's patrol tempo was deliberately moderated to preserve readiness. Operational histories of the conflict note that, while the presence of F-16s complicated India's aerial strategy (and likely dissuaded deeper airstrikes), the Vipers were never unleashed in a way that could have segued into a wider air war.

The F-16's role in the next chapter of Pakistani conflicts was one of transformation. Much like the Vipers in American service, Pakistani F-16s would evolve from high-altitude fighter-interceptors to low-flying precision tools of counterinsurgency. By the late 2000s, the Pakistani armed forces were confronting a new threat on their western flank. An insurgency had erupted in Pakistan's tribal areas—using caves, compounds, and the labyrinthine terrain to survive. Conventional air war doctrine thus gave way to the demands of counterinsurgency. Persistent surveillance, positive identification, and surgically-delivered airstrikes became the new normal. To that end, Pakistan began equipping select F-16s with advanced targeting packages (i.e. targeting pods, electro-optical/infrared systems, and laser designators). Each of these new accoutrements turned the F-16 into a precision strike asset.

The result was visible in 2009. During operations to clear militants from the Swat Valley and South Waziristan (notably Operation Rah-e-Nijat and follow-on campaigns), Pakistani F-16s flew strike missions against cave complexes, ammunition dumps, and rebel training camps. These airstrikes reportedly used 500-and 2,000-pound guided and unguided munitions, supported by new targeting-pod imagery, as well as intelligence shared from Allied assets. Contemporary stories from the *New York Times* noted that Pakistan was "injecting precision" into its air campaign, a capability that changed the tempo and effect of airstrikes within the tribal areas. Air raids became more surgical, and ground commanders increasingly relied on the F-16 to deny insurgents any sanctuary in the region.

Institutionally, this period of counterinsurgency rewired the PAF's tactical air power doctrines. The F-16's conversion into a multirole strike platform accelerated training for joint targeting procedures, sensor employment, and integrated mission planning. Squadrons that had previously flown high-altitude patrols were now mastering low-level surveillance orbits, targeting runs, and time-on-station tactics for optimal CAS engagements. Taken together, the new targeting suites and training priorities changed how Pakistani pilots approached the application of air power to the asymmetric battlefield.

By the time Operation Zarb-e-Azb launched in 2014, the architecture was in place: F-16s, armed with modern sensors and precision firepower, supported large-scale ground offensives that expelled the insurgents from the last of their safe havens. The public face of the campaign, however, was mixed. The operational success of counterinsurgency air power was colored by massive internal displacements and post-campaign reconstruction efforts. Militarily, however, it marked the final maturation of the Pakistani Viper. Under PAF control, the F-16 had become a lethal instrument in the realm of low-intensity, unconventional warfare.

The arc from Kargil to the western insurgency demonstrated the F-16's enduring strength: Adaptability. In Kargil, the aircraft helped deter and influence strategic decisions simply by its presence. During the tribal insurgency a decade later, the F-16 became a workhorse of precision airstrikes, reconnaissance, and air-ground integration. Both campaigns left their mark on the pilots, doctrines, and on the sense of what air power could achieve. For the PAF, the F-16's service over the western frontier wasn't a single moment, but a long journey of combining air power with restraint and purpose.

In February 2019, however, the long-standing rivalry between India and Pakistan once again erupted into armed conflict. This time, the spark came when a suicide bomber from Jaish-e-Mohammed killed more than forty Indian paramilitary troops in Kashmir. India retaliated with an airstrike by Mirage fighters on a suspected militant camp near Balakot. For the first time since 1971, Indian jets had crossed deep into Pakistani territory, and the die was cast for an aerial confrontation.

The next morning, Pakistan responded in kind.

During the predawn hours of February 26, the PAF launched Operation Swift Retort, a carefully-planned demonstration airstrike intended to show resolve without escalating into a full-scale war. According to Pakistani accounts, their F-16s and JF-17s crossed the Line of Control, dropping bombs into open areas near Indian troop positions in Rajouri and Poonch. All told, it was a pyrotechnic show-of-force meant to draw out the Indian Air Force (IAF).

As it turned out, the Pakistanis didn't have to wait long. Indian Sukhoi Su-30s, Mirage 2000s, and MiG-21s scrambled to intercept. What followed was a chaotic and fast-moving dogfight in the skies over Kashmir—the most significant air clash between the two nuclear-armed neighbors in more than forty years.

At the heart of the engagement were Pakistan's *Block 52* F-16s from the No. 11 and 19 Squadrons, armed with AIM-120 missiles. It was the first time in South Asia that a radar-guided, BVR missile would be unleashed in combat. According to US defense sources, at least one AMRAAM was fired that morning; while other sources claimed as many as four.

Pakistani claims, however, were more dramatic: The PAF announced that their F-16s had downed two Indian aircraft: a MiG-21 and Su-30. The Indian Air Force, in turn, acknowledged only the loss of a single MiG-21, flown by

Group Captain Abhinandan Varthaman, who ejected safely over Pakistani territory. Captured by villagers and then handed over to the Pakistani Army, his dignified behavior in captivity quickly turned him into a national hero in India. Bowing to international pressure, however, Pakistan repatriated him on March 1, 2019—a gesture that helped cool the crisis.

The alleged Su-30 shootdown, however, remains inconclusive. India denied losing any Sukhois that day, while Pakistan insisted that one of their F-16s had scored the kill. At this writing (summer 2025), there has been no conclusive evidence to confirm or deny the claim. Complicating matters further, India asserted that Abhinandan Varthaman had downed a Pakistani F-16 before his own MiG was lost—a claim which, likewise, remains unverified. In fact, subsequent US inspectors (all of whom were granted access to PAF bases) reportedly confirmed that all Pakistani F-16s were present and accounted for. The US report, published in *Foreign Policy* on April 4, 2019, concluded that there was no evidence of an F-16 having been shot down by enemy fire, much less by an antiquated MiG-21.

Amidst the haze of contradicting narratives, however, one fact was indisputable: Pakistani F-16s had proven their worth. Their AMRAAM missiles had forced the IAF to fight at a disadvantage, with several Su-30 pilots reportedly retreating under missile warnings. For New Delhi, it was a sobering reminder that the PAF's small but well-trained F-16 cadre could hold their own against larger Indian formations. In Pakistan, Swift Retort was celebrated as a victory of operational planning and execution. For India, however, the loss of Abhinandan's MiG-21 (and the lack of clear evidence to confirm

Pakistani losses) was a political and strategic "black eye."

But for both countries, the clash underscored the volatility of their ongoing rivalry.

In a matter of minutes, two nuclear powers had escalated the conflict from a low-level terrorist attack to a full-fledged aerial campaign.

And the F-16 had once again taken center stage. For Pakistan, it leveled the playing field in the face of India's numerical superiority. Across the border, it was a reminder that even legacy platforms like the MiG-21 could play decisive roles, albeit at a much greater risk. And to the international community, the skies over Kashmir were a chilling reminder that peace in the subcontinent was never a guarantee.

Turkey

From the rugged steppes of Anatolia to the waters of the Aegean, Turkey has long stood at the crossroads of empires, where the struggle between East and West has echoed for centuries. In the modern era, however, many of these long-standing rivalries have come under the scepter of the Fighting Falcon. In many ways, Turkish F-16s have become the face of air power in the Near East, projecting influence from the shores of the Aegean to the rugged mountains of Mesopotamia.

In the contested skies above the Aegean Sea, Turkish Falcons have tussled with Greek fighters over disputed airspace for decades. Over northern Iraq, Turkish F-16s struck at the heart of the Kurdistan Workers Party (PKK) in 2008—the latest episode in a campaign that has dragged on since the late 1970s. And in 2020, as Armenia and Azerbaijan once again clashed over the Nagorno-Karabakh region, allegations emerged that Turkish F-16s had flown in the service of Azerbaijan—phantom killers in a conflict where hard proof remains elusive.

For much of their respective histories, Greece and Turkey have feuded over their possessions in the Aegean Sea—islands, territorial waters, and even airspace. In the arena of territorial airspace, the heart of their dispute lies within their comparative concepts of distance. Greece asserts a 10-mile air defense zone off its coast, while Turkey recognizes only six miles.

Every year, this ambiguity leads to dozens of intercepts over the Aegean Sea, with both sides scrambling their fighters to shadow one another. By late 2019, data from the Hellenic Air Force showed more than 3,500 airspace "violations" by Turkish aircraft in just nine months, on pace to exceed 4,000 for the year.

Each incursion prompted an intercept by Greek F-16s or Mirage fighters.

Year after year, the cycle repeats itself, though weapons are rarely fired. As one Hellenic Air Force general explained: "We see them up there, fighting it out every day"—but these are mostly overflights and mock engagements.

In this ongoing game of brinksmanship, however, there have been a few deadly encounters.

In February 1995, for example, a Turkish F-16C crashed into the Aegean after being pursued by a pair of Greek Mirages. The pilot ejected safely, and the mishap has been attributed to either fuel depletion or mechanical failures precipitated by the duress of trying to escape.

On October 8, 1996, during the Imia/Kardak Island Crisis, two Turkish F-16s alongside four F-4 Phantoms were on a SEAD training mission over the Aegean when they "violated" Greek airspace north of Chios. A pair of Greek Mirage fighters (one flown by Captain Thanos Grivas) were vectored to intercept. The Turkish flight allegedly refused to leave Greek airspace, claiming that they were flying over international waters.

Thus, Captain Grivas readied his Mirage for a missile engagement.

During the brief dogfight that ensued, Grivas skillfully downed the Turkish F-16D with an R.550 missile shot. Of the two-man Turkish crew, only the Weapons System Officer, Lieutenant Colonel Osman Cicekli, survived the shootdown. He ejected safely into the Aegean. But for reasons unknown, the pilot, Captain Nail Erdogan, failed to eject and purportedly died in the crash.

In the aftermath of the incident, both the Turkish and Hellenic Navy conducted an extensive search of the area where Erdogan's F-16 purportedly went down. Sadly, Erdogan's remains were never recovered.

For many years thereafter, neither side spoke at length about the incident. In fact, the F-16 shootdown was widely held to be a rumor until 2003. That year, a former Turkish naval commander confirmed that a Greek warplane had downed a Turkish F-16 over the Aegean in 1996. Years later, Turkish Defense Minister Ismet Yilmaz reiterated that confirmation, adding that the Turkish F-16 had indeed been downed by a Greek Mirage.

A decade later, on May 23, 2006, a Turkish and Greek F-16 rammed into each other over Karpathos Island. Plunging from more than 26,000 feet, both pilots ejected. The Turkish pilot, Lieutenant Halil Ozdemir, safely parachuted into the sea and was rescued by a passing ship. The Greek pilot, Captain Kostas Iliakis, died during the descent and his body was later recovered at sea. As expected, each side blamed the other. Turkey said the mid-air collision occurred over international waters, while Greece said the encounter happened over their own territory.

Beyond these clashes, the two air forces still engage each other in spectacular (but non-lethal) shows of force. Video footage has shown F-16s on either side diving and weaving in mock combat, with both sides taking care not to fire live ammunition.

Still, the loss of a pilot lingers heavily against the backdrop of this high-stakes rivalry. Observers have noted that Greek F-16s scramble to intercept Turkish flights almost daily during periods of intense diplomacy and political brinksmanship. The Aegean has thus become a *de facto* battlefield where Turkish and Greek F-16s test each other's resolve in a series of aerial duels. It remains a high-speed drama that, as history shows, can turn lethal at any moment.

Farther south from the Aegean Sea, Turkish F-16s became the spearhead of Ankara's long and bitter struggle against the PKK, which had entrenched itself in the mountains of northern Iraq. What began with sporadic air raids in the 1990s slowly evolved into a crescendo of air power by the mid-2010s—intensifying by 2018 into something more organized, more systematic, and unmistakably more lethal. Out of this transformation emerged the series of offensives which Turkey branded the "Claw" operations. It would become a continuous chain of airstrikes tightly synchronized with Turkish Army ground operations.

The new tempo was nothing short of dramatic. Indeed, defense analysts observed that Turkey's modern air campaign had nearly *tripled* the effects of its counterinsurgency effort, shifting the battlefield deep into Iraqi territory. And, as expected, the F-16 took center stage during the opening rounds: Thundering across the border to eliminate bunkers, cave complexes, and staging areas before Special Operations teams and helicopters pressed in behind them.

But even before the Claw series had been formally christened, there were harbingers of what was to come. In June 2016, Turkish warplanes (frequently a mix of F-16s and F-4s) swept across the frontier to pummel priority PKK targets. Reports from that surge spoke of weapons caches and shelters torn apart by precision airstrikes. It was a prelude, of sorts, to the campaigns that would soon follow.

That moment arrived in June 2020 with the launch of *Claw-Eagle*, the first major air power-led offensive of the Claw series. F-16s cut through the skies over Hakurk and Qandil, their bombs hammering guerrilla bases hidden among the forests and mountains. A few days later came *Claw-Tiger*, shifting the fight to the Haftanin region. The Ministry of Defense boasted that more than 150 PKK targets had been destroyed, while Turkish Special Forces (ferried in via helicopter) descended into the valleys as F-16s roared overhead.

But the escalation didn't stop there.

In April 2021, President Recep Erdogan declared the beginning of *Claw-Lightning* in the Metina region, vowing to "end the presence of the terror threat along our southern borders." The scenes that followed lived up to the rhetoric. Videos released by the Ministry of Defense showed Turkish fighters screaming low over the ridgelines, trails of missiles cutting into hidden PKK strongholds, while Turk infantrymen descended into the remote valleys. *Claw-Thunderbolt* began in parallel, extending the fight into the Zap region, amplifying pressure on the PKK's northern redoubts.

Together, these operations became a relentless drumbeat of rapid-fire air power. Reports described F-16s flying near-continuous sorties between April and June 2021, delivering stand-off precision firepower into the mountainous network of tunnels and fortified caves. Turkish communiqués tallied hundreds of "neutralized" insurgents, while the PKK sheepishly admitted to their own heavy losses. The Iraqi government, meanwhile, protested that the airstrikes violated their national sovereignty; but Turkey stood firm, insisting that the anti-PKK raids were a matter of self-defense.

On Turkish television, the images were triumphant: F-16s streaking across the jagged skylines, commandos raising flags over seized terrain, and UAVs circling overhead. To the Turkish public, the Claw campaigns had become a symbol of strength, perseverance, and justice rendered against an implacable foe. To the Kurds in northern Iraq, however, the Claw offensives were a brutal reminder of Turkish military might.

As Turkish F-16s thundered across the border into Iraq, another storm was brewing to the northeast. In September 2020, the long-disputed enclave of Nagorno-Karabakh once again became a flashpoint between Armenia and Azerbaijan. Turkey's allegiance was no secret: They stood firmly on the side of Azerbaijan, urging the government in Baku to "take matters into its own hands."

Yet when it came to direct military involvement, Turkish leaders were more cautious. They claimed solidarity, but denied sending combat aircraft into the fight. International monitors seemed to agree: US and Russian radar sweeps reported no unfamiliar aircraft in the skies over Armenia. French President Emmanuel Macron, meanwhile, decried Turkey's "warlike" rhetoric but admitted there was no proof of F-16 involvement.

Still, on September 29, 2020, Armenia claimed one of its Su-25s had been shot down by a Turkish F-16, killing pilot Major Valeri Danelin. Photos of the wreckage circulated throughout the media, but both Turkey and Azerbaijan denied the allegations. Turkey's Ministry of Foreign Affairs claimed the allegations were "absolutely untrue," a denial echoed in international press coverage.

Independent scrutiny never confirmed the claim. No radar tracks, satellite imagery, or credible video footage ever placed a Turkish F-16 in action over Nagorno-Karabakh. Open-source investigators documented extensive use of Turkish-built Bayraktar TB2 drones, but nothing to tie Turkish Vipers to the fighting.

What is known, however, is that Turkish F-16s were stationed in Ganja, Azerbaijan, during joint exercises just before the war—and they remained there during the conflict. But whether they were meant as a deterrent or held in reserve remains uncertain.

In the end, the clash over Nagorno-Karabakh underscored a new reality of modern warfare: The battle for narrative runs alongside the battle for territory. Rumors of F-16 air cover spread quickly, but in the absence of proof, Turkish F-16 involvement remains unverified—only rumors, allegations, and denials exist. What lingers instead is a cautionary tale: In an age where rumors can carry strategic weight, even the unsubstantiated *hint* of an F-16 can shape the public's perception of war.

Venezuela and Thailand

Throughout its illustrious history, the F-16 has often found itself performing missions far beyond what its Cold War designers had anticipated: Internal security and crisis management; border disputes; counterinsurgency; law enforcement; and the shadowy wars of the modern age. Nowhere is this clearer than in the turbulent stories of Venezuela and Thailand. From the Caribbean basin to the jungled frontiers of Southeast Asia, the F-16 has been a weapon of political survival and, at times, a participant in battles where the line between warfare and law enforcement is heavily blurred.

When Venezuela acquired its first F-16s in 1983, the purchase marked a major milestone: Caracas became the first Latin American government to receive the Fighting Falcon. For the Venezuelan Air Force, the F-16 carried an aura of prestige—a message that Venezuela would not be overshadowed by regional rivals or Cuban-aligned insurgencies simmering in Central America. For years, these Falcons flew routine fighter patrols over the Caribbean skies. By the early 1990s, however, they would be thrust into their country's own internal drama.

On November 27, 1992, dissident elements of the Venezuelan Air Force and Navy launched a second coup attempt against President Carlos Andrés Pérez. Unlike the failed uprising from the previous February, the new rebellion relied heavily on air power. Rebel crews scrambled OV-10 Broncos, AT-27 Tucanos, and even Mirage fighters to strike loyalist targets, leaving government forces scrambling to respond. Over the capital, Venezuelans looked up in shock as their own jets thundered overhead—some flown by mutineers, others by loyalists defending the constitutional order.

The most dramatic moment came when loyalist F-16s squared off against rebel aircraft in the skies near Caracas. Two F-16s, piloted by Captain Guillermo Beltran-Vielma and Lieutenant Helimenas Labarca, took to the skies with a bare-bones mission load: Each plane carrying an extra fuel tank, defensive flares, and only 280 rounds of 20mm ammunition—with no air-to-air missiles on board. Following orders to eliminate any insurgent aircraft, the pilots turned their Vipers loose over central Venezuela. Near Yaritagua (between Caracas and Maracay), Beltran-Vielma locked onto a rebel OV-10 Bronco at low altitude. A short 20mm burst into the OV-10's twin engine sent its rebel pilot, Lieutenant Rodolfo Domador, into a fatal crash. Domador ejected, but later died from injuries sustained during the descent. The second rebel Bronco, piloted by Air Force dissident Carlos García, attempted a bold evasive loop, but Beltran-Vielma sheared off one of the Bronco's wings with another multi-round burst, forcing Garcia to eject.

Meanwhile, Lieutenant Labarca, piloting the other F-16, refueled at Barquisimeto and vectored back towards Maracay, where rebel AT-27 Tucano attack planes were preparing to strike loyalist ground forces. Labarca thundered in from above, strafing two of the rebel Tucanos as they were taxiing from the airfield. One Tucano pilot, Luis Berroterán, was badly hit and leapt from his cockpit. The second Tucano flyer, Ali Nicolacci, was mortally wounded on the runway, succumbing to his injuries shortly thereafter.

As loyalist forces regained the upper hand, regime-piloted F-16s also conducted airstrikes against key rebel positions, including the rebel-held base at La Guaira. Precision bombing runs helped neutralize the remaining insurgent air assets, turning the tide in favor of the government.

By mid-morning, the loyalist F-16s had blunted the rebels' air power advantage, allowing loyalist ground forces to regain the initiative. In an ironic twist of fate, heavy rainfall stalled the fighting and, by midday, the coup had collapsed. Nearly 100 insurgents fled aboard a commandeered C-130 to Peru, while those who remained were rounded up.

Captain Beltran-Vielma and Lieutenant Labarca, meanwhile, were hailed as heroes. For the first time in history, Venezuelan Vipers had been pointed not at foreign adversaries, but against fellow countrymen. Yet, their intervention (both in the air and against rebel-held positions on the ground) proved decisive in saving the government of Carlos Andres Perez, cementing the F-16's reputation as both a weapon of national prestige and a guardian of the embattled state.

By the dawn of the 21st Century, the Venezuelan Vipers faced a different kind of war: A war without battlefronts, fought in shadowy skies against non-conventional enemies with no political agenda.

This time, theirs was a war against the *traficantes.*

Indeed, the rise of drug trafficking across the Caribbean had turned Venezuela's airspace into a corridor for drug flights in and out of Colombia. Smugglers in fast aircraft (often business jets or twin-engine turboprops) used the vast, unpoliced interiors and unguarded coastlines as staging areas. In response, Venezuelan F-16s became the tip of the spear in combating narco-terrorism.

One dramatic case unfolded in July 2024. On that day, local radar detected an unknown aircraft entering

Venezuelan airspace. Despite repeated attempts by air traffic control to establish contact, the aircraft—identified as a Piper PA-34-200T Seneca II—continued on its course, disregarding all directives from ground control.

Moments later, the Venezuelan Air Force scrambled a pair of F-16s, tasked with intercepting the intruder and, if necessary, eliminating it. Upon reaching the wayward Piper, the lead F-16 pilot issued several warnings and ordered the intruder to land at a designated airport. However, the intruding aircraft ignored the commands and began attempting evasive maneuvers.

As the situation escalated, the twin-engine Piper executed a forced landing in a field near Turén. Upon inspecting the wreckage, authorities discovered the plane had been registered in Brazil. Amongst the wreckage, investigators also found the remains of the pilot and documentation linking the aircraft to drug trafficking activities. Notably, a Mexican passport and a US-issued flight license were among the recovered items, further corroborating the aircraft's involvement in illicit operations.

These events underscored Venezuela's commitment to safeguarding its territorial sovereignty and combating the drug trade. Through a combination of aerial intercepts and targeted ground operations, the Venezuelan Air Force continues to assert control over its domestic airspace and disrupt illegal activities that threaten national security. For Caracas, the message is clear: The F-16 is an instrument of state sovereignty, used to enforce laws in the grey zone between policework and military conflict.

Still, since the late 2010s and early 2020s, Venezuela's F-16 fleet has fallen on hard times. Spare parts have grown scarcer under economic sanctions. Maintenance programs have lapsed and, today, only a fraction of the original F-16 force remains airworthy. But the legend of the Venezuelan Vipers has endured, remembered both for the rebellions they've crushed...and the drug flights they've destroyed.

Halfway across the world, Thailand's acquisition of the F-16 reflected similar ambitions. Introduced in the late 1980s, the Thai Falcons were meant to secure the kingdom's skies against regional threats and to modernize an air force that had long relied on older American and European aircraft. As such, the F-16s soon proved themselves not only as deterrents, but as aerial vanguards in the regional border conflicts.

The most volatile of these border disputes flared with the neighboring Kingdom of Cambodia. The Thai-Cambodian frontier, dotted with ancient temples and forested ridges, has long been a source of contention between the two monarchies. In July 2025, tensions erupted once again, this time near the Ta Moan Thom and Preah Vihear temple complexes, escalating into the most severe conflict between the two kingdoms in more than a decade.

On July 24, in response to artillery attacks from Cambodian forces that resulted in civilian casualties, the Royal Thai Air Force (RTAF) scrambled six F-16s from Ubon Ratchathani Air Base. RTAF Vipers targeted Cambodian military positions near the disputed Ta Moan Thom temple complex.

Following the airstrikes, however, Cambodian state media and local outlets claimed that their air defense forces had downed a Thai F-16. However, the RTAF denied these allegations, stating that all six aircraft completed their missions and returned safely.

Meanwhile, the conflict intensified rapidly on the ground, with both sides exchanging heavy artillery and rocket fire across the 500-mile border. At least twenty were killed, and more than 130,000 civilians were displaced due to the violence. Thailand's Ministry of Foreign Affairs accused Cambodia of deliberately attacking civilians, using rocket artillery to target populated areas. The UN Security Council, for its part, convened to discuss the situation, while the ASEAN members brokered a ceasefire agreement. Despite these initial agreements, both sides have accused each other of violating the ceasefire, and tensions remained high.

Nevertheless, the July 2025 airstrikes marked a significant chapter in the Thai-Cambodian border conflict, highlighting the role of modern air power in regional disputes. While the immediate threat of a protracted, high-intensity war has been averted, the underlying territorial issues remain unresolved.

Venezuela and Thailand could hardly be more different: One is a Latin American republic beset by coups and economic collapse, the other a Southeast Asian monarchy balancing tradition with modernity. Yet under both flags, the F-16 became an indelible instrument of state power. In Venezuela, it was the fighter that defended democracy against rebel forces. Then, it became a frontline vanguard in the War on Drugs. In Thailand, it was the "eye in the sky" over contested borders, projecting air power across the jungles and temples of the Mekong frontier.

Their stories, like many others where the F-16 has played a central role, illustrate the Viper's unyielding flexibility. Whether in the service of banana republics or the monarchies of Indochina, this lightweight Cold War fighter has proven itself capable of serving as a coup-breaker, anti-drug enforcer, and border guardian alike. The saga of Thai and Venezuelan F-16s, therefore, is more than just a chapter in aviation history. It is a reminder that aircraft are never just machines. They are instruments of political willpower.

Ukraine

April 12, 2025: In the quiet expanse of Ukrainian airspace, a lone F-16 thundered across the sky. At the controls was Captain Pavlo Ivanov, a 26-year-old Ukrainian Air Force pilot. For months he had trained relentlessly, learning the intricacies of this American-built fighter—a machine once unimaginable in Ukrainian hands. To Ivanov, the cockpit was no mere collection of gauges and switches. It was a lifeline in the ongoing war against Russia.

Few could have predicted such a scene prior to the war: A former Soviet republic, long reliant on legacy MiGs, was now flying a Western-built fighter into combat against the very air force that once defined the Soviet empire. Yet, here it was: The F-16 as a symbol of resistance against Russian air power. For the Ukrainian Air Force, the F-16 embodied more than just firepower; it embodied the will of a nation fighting for survival. Still, Ivanov was well aware of the lingering, unpleasant truth: No machine, regardless of its technological edge, could erase the uncertainty of combat.

When the first F-16s arrived in Ukraine during the summer of 2024, it was a major geopolitical shift. For decades, Kyiv had maintained an aging fleet of Soviet airframes—

including the MiG-29 and Su-27. Given their age and comparative capabilities, however, these Glasnost-era jets were able to harass, but rarely challenge, the Russian Air Force. But with the F-16, Ukrainian pilots could now stand toe-to-toe against the latest 4th-Generation+ fighters. Advanced radar, precision firepower, and state-of-the-art targeting systems—these were the tools that Ukraine's air force had never possessed until now.

The F-16's delivery to Ukraine was no accident. Indeed, it was a calculated move by Washington and its NATO allies, born of hard geopolitics. On February 22, 2022, the Russian Armed Forces had launched a full-scale invasion of Ukraine. Vladimir Putin, the President of Russia, had long harbored ideations to annex the former Soviet republic. But whatever his motivations may have been, Putin expected his air and ground forces to make short work of the Ukrainian resistance. Military experts and media analysts predicted the total collapse of Ukrainian defenses within a few weeks to a few months.

But even the most seasoned analysts failed to anticipate the ferocity of Ukraine's resistance.

The first wave of the air campaign began at 5:00 AM (local time) on February 24, 2022. A total of 75 Russian aircraft were committed to the initial onslaught, accompanied by a fusillade of land-based ballistic and naval-based cruise missiles. Early battle damage assessments confirmed more than 80 Ukrainian targets had been destroyed, including eleven airfields. A near-concurrent cyber-attack, meanwhile, crippled many of Ukraine's Command & Control nodes.

The Ukrainians, though shaken by the multi-pronged attack, fought back valiantly. But to the Western world, Moscow's aggression was a threat to regional security and the fragile balance of power in Europe. Fears persisted that NATO might be drawn into the conflict, precipitating the dreaded "World War III" scenario that every Cold War strategist had sought to avoid.

But passive silence wasn't an option, either.

Come what may, Western leaders felt they had to do *something*. Economic sanctions soon followed, while some former Eastern Bloc partners donated their excess MiG-29s to the Ukrainian Air Force. The US, however, shocked the world when it delivered the AGM-88 HARM missiles to Ukraine in late 2022.

Although it had a reputation for being one of the best air-to-ground missiles in the world (due in no small part to its service in Afghanistan and the Balkans), the AGM-88's arrival in Ukraine was perplexing because the system wasn't compatible with any Soviet-built aircraft. Indeed, none of the MiG or Sukhoi wing pylons could accommodate the missile fitment; and the AGM-88 could only be operated through a digital interface.

Nevertheless, Ukrainian aerospace engineers and American contractors developed a solution to integrate the NATO-spec missile into the MiG-29's airframe. They installed specially-modified wing pylons and equipped the pilot with an iPad in the cockpit to operate the HARMs in combat.

US Undersecretary of Defense Dr. William LaPlante confirmed in 2024 that the Ukrainian Air Force had implemented these adaptive measures. "Working with the Ukrainians," he said, "we've been able to take many Western weapons and get them to work on their aircraft where it's basically controlled by an iPad by the pilot. And they're flying it in [the] conflict like a week after we get it to him." After integrating the AGM-88 HARMs, Ukrainian

MiG-29s also began using the Joint Direct Attack Munition-Extended Range (JDAM-ER) precision-guided bombs.

But even with the arrival of these new air-to-ground munitions, the MiG-29 was still at a disadvantage considering the limitations of its organic fire controls. Because the Fulcrum could not accommodate the AN/ASQ-213 targeting system, the pilot had to fly much closer to his intended target before launching the missile. By contrast, the AN/ASQ-213 would allow the pilot to track the location of hostile radar systems and engage targets from greater standoff distances, thereby minimizing exposure to enemy air defense batteries.

After the high-profile downing of a MiG-29 during a strike mission in 2022, Ukrainian defense analysts sounded a clarion call: Only the F-16, with its coveted AN/ASQ-213 and proven mastery of SEAD warfare, could tip the balance in the skies. At the time, their words seemed aspirational. Yet their prediction would soon become reality. Two years later, the first F-16s arrived in Ukraine, gifted from Denmark, Belgium, and the Netherlands.

Arming Kyiv with the F-16 sent a clear message to the leaders of Eurasia: NATO was prepared to defend its eastern flank with deadly force; and Russian jets would not go unchallenged in the skies over Ukraine.

Yet the promise was not without its perils.

True, the F-16s had given Ukraine an extra dose of lethality, but their very presence made them a high-value target. Russian air crews relentlessly scoured the Ukrainian countryside for any sign of an F-16 base. Ukraine's operational security and secrecy had shrouded their exact numbers and locations. Still, the eagerness, diligence, and near-desperation of the Russian Air Force to find an F-16 base was both significant and humorously ironic. The symbolism was inescapable: Every Viper was a threat to the narrative of Russian air supremacy.

But the F-16's journey from delivery to viability was long and tumultuous. For pilots raised on the MiG-29, their acclimatization to Western aircraft was nothing short of revolutionary. Under tutelage from the 162nd Wing—America's premier unit for training international Viper crews—at Arizona's Morris Air National Guard Base, Ukrainian pilots immersed themselves in a training regimen that demanded not just aerial skill, but a new way of thinking.

Classroom hours stretched late into the night. Simulators threw them into dogfights against Su-35s, SAM launches, and EW attacks. English-language technical manuals replaced Cyrillic checklists. NATO doctrines replaced years of Soviet-era instincts. Every pilot logged at least 90 hours in the cockpit, a crucible that tested their ability to survive the gamut of air combat scenarios.

For the pilots themselves, the F-16 was agile and responsive in ways the MiGs never were. Its radar could track targets invisible to the old systems; and its AMRAAMs and HARMs could reach ranges unthinkable to Soviet arsenals. But mastery also demanded humility. Seasoned combat veterans found themselves students once again. But through sweat, discipline, and perseverance, the transformation from Eastern Bloc metrics to NATO-style airmanship took hold. The program forged not just pilots, but warriors of a new tradition: Aviators who could think, fight, and engage the enemy with NATO-level precision.

Training, however, was theory. War was something else entirely. When the first Ukrainian F-16s entered combat, they faced an enemy that was merciless and numerically superior. Russian interceptors stalked the skies while SAM batteries crisscrossed the occupied territories. Each mission would become a test of nerve as much as skill.

On the eve of their combat debut, these first-generation Ukrainian F-16s were configured with AIM-9s, AIM-120s, and a handful of ground-attack munitions as needed. Although it was a moderate weapons suite by 21st Century NATO standards, it nevertheless gave Ukraine an enormous boost of firepower over their rapidly depleting fleet of MiG-29s. Early on, US Air Force specialists reprogrammed the F-16s' EW systems with threat libraries tailored to counter Russian-built threats. Taken together, these two strands—the NATO-spec ammunition and a freshly-tuned EW suite—would mark the F-16's baptism by fire in the skies over Ukraine.

Ukrainian Vipers flew into combat on August 26, 2024, during one of the largest Russian air offensives of the war to that date. F-16s scrambled into a hot sky, thick with incoming cruise missiles and one-way attack drones. Despite their modest combat loads, the Vipers' onboard sensors and EW software gave them an unexpected edge. By confusing some seeker heads and hardening their jets' own signatures, the US reprogramming effort helped the F-16 survive in an environment where Russian heat-seekers and SAMs stalked the third dimension with lethal efficiency.

On that morning of August 26, one name rose quickly into legend: Lieutenant Colonel Oleksii Mes, a decorated fighter pilot who had been a leading advocate for bringing the F-16s into Ukrainian service. Accounts from Ukrainian authorities report that Mes and his comrades intercepted a wave of incoming cruise missiles that day. Official briefings credited the F-16s with downing multiple warheads—missile-on-missile shots oddly reminiscent of the Patriot-Scud showdowns from the Gulf War.

Throughout the mission, Mes's radio transmissions carried the clipped coolness of a well-trained pilot: Precise, business-like radio calls as he engaged multiple threats. He had helped popularize the view of the F-16 as a "Swiss Army knife" of the sky. And he used every tool the jet afforded him that day: Long-range missiles against some threats, infrared missiles against others, and when circumstances narrowed, the 20mm cannon to finish off low-flying targets.

Sadly, his mission that day ended in tragedy. Mes's aircraft was purportedly downed by a friendly SAM, and the Ukrainian Air Force sullenly confirmed his death as their first F-16 combat loss. The shootdown prompted a thorough investigation (which remains open as of this writing) but it remains a sobering reminder of the price paid when new equipment is thrown into a brutal air war.

These early months of the Viper in Ukraine read like a case study in extremes. On the one hand, there were brilliant success stories, achieved by pilots who were still learning. On the other hand, there were equally stark losses as Ukrainian crews pushed the F-16 to its limits. On December 13, 2024, for example, during a massive Russian airstrike involving more than 90 missiles and nearly 200 attack drones, a Ukrainian F-16 pilot achieved a feat that his commanders would later call "historic." He downed *six* Russian cruise missiles in a single sortie. Ukrainian officials announced the claim publicly, noting that the pilot had used the F-16's 20mm autocannon to destroy the missiles that had survived his earlier AIM-9 and AIM-120 shots. It was a hair-raising tactic that required him to close within distances far shorter than the optimal range for engaging high-speed cruise weapons.

However, these moments of tactical brilliance could never mitigate the overarching strategic danger. Russia's layered air-defense networks (including SA-21 launchers, shorter-range mobile SAM batteries, and BVR air-to-air missiles) meant that every sortie was a calculated risk.

Such was the case when Captain Pavlo Ivanov went airborne on April 12, 2025. The proud 26-year-old pilot had just completed his F-16 training when his Viper was shot down near Sumy. Reports indicate that Ivanov was likely struck by a Russian SA-21, but the precise chain of events remains the subject of an interagency investigation. What is not disputed, however, is the human cost: Ivanov's death showed that even the most capable aircraft offered no immunity against modern, integrated air defense networks.

As spring turned to summer, losses continued to mount. On May 16, 2025, Ukraine confirmed that another F-16 was lost after an "emergency situation" during a mission to repel a Russian drone attack. The pilot safely ejected and was recovered. Early briefings described a harrowing sequence in which the pilot engaged multiple aerial threats, downing several before an unexplained "onboard emergency" had forced his ejection. Still, the incident highlighted an unpleasant reality. Flying and maintaining a small fleet of Western jets in a high-intensity air war was no easy feat. Logistics and sustainment were the thin margin between victory and catastrophe.

Then, on June 29, 2025, Ukraine suffered one of its most heart-wrenching losses. A massive Russian nighttime raid (including hundreds of drones and dozens of cruise and ballistic missiles) erupted into the sky. Lieutenant Colonel Maksym Ustymenko's F-16 scrambled to meet the threat.

Ukrainian briefings described Ustymenko as exhausting "all of his onboard weapons" to stop incoming drones and missiles. Commanders said he had shot down seven aerial targets before his aircraft suffered critical damage, apparently from debris or a weapon impact, and he was unable to recover. He died in the resulting crash. President Zelensky posthumously awarded him the title Hero of Ukraine. The scene read like a proverbial "last stand" from a Hollywood script: A pilot using every tool at his disposal, trading standoff safety to protect civilians below, and finally paying the ultimate price.

Taken together, these episodes mapped a story arc that was both tactical and groundbreaking. The F-16's arrival had revamped Ukrainian air doctrine. The US-led effort to tune EW suites for Russian threats, along with the early success stories (i.e. missile/drone intercepts, high-risk SEAD missions, and precision airstrikes) all demonstrated what the Vipers could do.

Yet their attrition also exposed their limits.
A handful of Western jets couldn't single-handedly erase a tightly-networked air defense grid.

Still, the geopolitical calculus had shifted irreversibly. With the F-16, Ukraine could contest the skies in ways once thought unimaginable. Precision airstrikes disrupted Russian formations; high-speed intercepts their blunted missile attacks; and precision firepower obliterated their command posts far behind enemy lines.

For Kyiv, the Viper was more than just a tool. It was proof that Ukraine was no longer chained to its Soviet past. Interoperability with NATO air forces was no longer an aspiration; it was a demonstrable *fact*. Each sortie brought Ukraine closer to a military doctrine shared with its Western allies; and each pilot graduating from the F-16 conversion course was a thread that bound Kyiv tighter into the security architecture of mainland Europe.

But the geopolitical implications have gone far beyond tactics and equipment. Air parity (if not air superiority) has been a strong bargaining chip at the negotiating table, a signal to both friend and foe. To Ukraine and her allies, it's been a clear sign of resilience and determination. To Russia, it's been a whisper of deterrence.

Yet through all the grand strategy, one truth has remained: The *machine* cannot be separated from the *man*. Just as it has done in Western air forces, the F-16 has, in many ways, become an extension of the Ukrainian airmen who've flown and maintained it. In the skies over Kyiv and Kharkiv, each of the fallen pilots have embodied the paradox of modern warfare, blending ancient courage with futuristic steel. Their triumphs and tragedies reveal that the path to air supremacy isn't paved by technology alone, but in the resilience of those who dare to fly.

At this writing (summer 2025), the War in Ukraine is still ongoing. But as the war approaches its fourth anniversary, it's clear that the F-16 has become more than just a stopgap for the Ukrainian Air Force. It has, in many ways, become the new foundation. Indeed, the training, tactics, and sacrifices have precipitated a new doctrine of air power—a doctrine designed to *win the war* (or at least survive it) and *shape the peace* that follows.

For Ukraine and its armed forces, the journey has been perilous, yet encouraging. The former Soviet republic has stepped out from the shadows of its past and into the hard light of the future. But in that light, the silhouette of the F-16 has emerged as a symbol of defiance, resilience, and postwar hope.

Norwegian F-16AMs from the 331st Squadron in flight during Exercise Arctic Challenge 2019. Although Norway has since replaced their F-16s with the newer F-35, the Fighting Falcon had a long operational history with the Royal Norwegian Air Force. As a NATO member, Norway's F-16s served in the Balkans, Afghanistan, and during the 2011 intervention in Libya. *Alan Wilson*

Epilogue:
The Viper's Enduring Legacy

For nearly half a century, the F-16 Fighting Falcon has ruled the skies—not as the largest, the fastest, or even the most advanced jet of its kind—but as one of the most balanced, battle-proven, and enduring warplanes ever built. Born from the crucible of the Cold War, baptized by fire in the Middle East, and seasoned in the grey zone conflicts from Yugoslavia to the outer Caribbean, the F-16 has become the embodiment of agility, versatility, and the indomitable spirit of the fighter pilot.

When the Viper entered operational service in 1979, few could have predicted how profoundly this single-engine jet would reshape the face of modern air combat. Conceived as a "lightweight" companion to the F-15 Eagle, the F-16 was designed for one thing above all else: *Maneuverability*. The F-16 could outclimb, out-turn, and outfight nearly any other plane from its generation. And in the decades that followed, the F-16 would prove that agility, not brute strength, was the key to air superiority.

In the skies over the Bekaa Valley, Israeli F-16s annihilated Arab MiGs in a series of engagements that confirmed what the jet's designers had always believed: Technology, when fused with skill, could rewrite the rules of aerial combat. During the air campaigns over Lebanon in 1981-82, Israeli F-16s claimed dozens of aerial victories without a single loss.

And the legend of the Falcon was born.
In the years that followed, the F-16 became a global

phenomenon. America's allies embraced it not merely as a fighter jet, but as a statement of trust and partnership. From Europe to Asia, from the Middle East to South America, more than two dozen air forces adopted the Viper as the backbone of their tactical fighter fleets. It flew under the banners of NATO and the United Nations, striking hard when diplomacy failed, and guarding the peace that came thereafter.

In conflicts big and small, the F-16's combat record reads like a map of modern history. American Vipers struck deep into Iraq during Desert Storm, silencing enemy air defenses and securing air superiority over the desert. They returned to the same skies a decade later during Operation Iraqi Freedom. Over the Balkans, they hunted MiGs and shredded enemy radar sites with deadly precision. Turkish and Greek F-16s sparred across the Aegean Sea in their recurring peacetime duels. Pakistani Vipers faced Indian MiGs in moments of geopolitical brinkmanship. And in the Levant, F-16s from Israel, Jordan, and the UAE carried out pinpoint strikes against terrorist networks, often deep behind enemy lines.

Each of these campaigns told the same story: The F-16 was not merely an instrument of war, but a living testament to adaptability. Its flexible design facilitated re-inventions and innovations, which guaranteed its relevance long after newer jets had appeared. The early *Block 10* and *Block 15* Vipers, lean and stripped-down, gave way to the multirole *Block 50*s and *Block 60*s with advanced sensors, precision firepower, and digitalized cockpits. The same airframe that had once fired Sidewinder missiles at enemy MiGs could now deliver GPS-guided bombs onto moving targets from 20,000 feet.

But the Falcon's greatest legacy lies in the people who

flew it. Indeed, to fly the F-16 was to join a brotherhood that transcended borders. American, Israeli, Belgian, Danish, Pakistani, Polish, and Ukrainian pilots all spoke the same language of combat aviation. In their hands, the F-16 also became a canvas for national identity. Each air force painted its own story across the Falcon's wings: The desert camouflage of the Israeli Netz, the tiger-striped tails of NATO squadrons, or the patriotic color schemes of the USAF Thunderbirds slicing across the peacetime sky. Whether screaming low through canyon walls or rolling in tight formations before a cheering crowd, the F-16 had inspired not through spectacle alone, but through the quiet confidence that came from decades of combat excellence.

By the mid-2000s, many believed that the Viper's time had passed. Newer designs (including the F-22 Raptor and F-35 Lightning) seemed poised to take its place.

Yet the F-16 endured.

Nations continued to upgrade and modernize their fleets. Newer variants rolled off production lines in South Carolina, South Korea, and beyond. In Ukraine and other emerging air forces, the Viper is proof that even in an age of drones and stealth technology, the *human* pilot still carries the weight of homeland defense. In many ways, the F-16's endurance mirrors the story of those who flew it. It was never a perfect machine. It demanded skill, discipline, and constant respect from its pilots. But the F-16 rewarded those who mastered it. And as the years passed, the Viper came to embody a truth as old as military aviation itself: Courage, adaptability, and willpower are the timeless weapons of airmen everywhere.

Today, as it flies into its fifth decade of service, the F-16 is more than just a relic of the Cold War. It's a bridge between eras—a testament to the enduring relevance of human flight in a world increasingly run by automation. From its first dogfights over the Levant to its latest aerial battles in Ukraine, the F-16 Fighting Falcon is a reminder that innovation and spirit do not fade with time. The F-16's story, therefore, is a tale of endurance and transformation. It's the story of a lightweight fighter that refused to quit. It's the story of a machine that adapted as the world changed.

And so, the F-16 Fighting Falcon continues to fly, carrying the legacy of yesterday's battles...and the promise of those yet to come.

Select Bibliography

Archival Holdings:

Air Force Historical Research Agency
The Israel Defense Forces & Defense Establishment Archives
Library of Congress - Veterans History Project Collection
Virginia Military Institute – Military Oral History Archives

Primary Sources:

"366th Fighter Wing History: 1991–2002." *366th Gunfighters Association*, 2023.

388th Fighter Wing Public Affairs. "Vipers of '91: Hill's F-16s at War." 388th Fighter Wing.

Air Combat Command. "Killer Scouts and Night Vipers." ACC News Archive.

Clancy, Tom, and Chuck Horner. *Every Man a Tiger*. New York: Putnam, 1999.

Cohen, Eliezer "Cheetah". *Israel's Best Defense: The First Full Story of the Israeli Air Force*. London: Orion Books, 1993.

Dutch Ministry of Defence. *The Use of Air Power over Bosnia, Croatia, and Kosovo: Dutch Contribution, 1993–2001*. The Hague, Netherlands: Government Printing Office, 2004.

Halperin, Merav, and Aharon Lapidot. *G-Suit: Combat Reports from Israel's Air War*. Sphere Books Ltd., 1990.

Hampton, Dan. "The Weasels at War." *Air & Space Forces Magazine*, July 1991.

Head, William, and James Tindle. *Operation ALLIED FORCE: Special Study 19-02*. Air Force Materiel Command, HQ AFMC History Office, 2019.

Hill Air Force Base News. "Hill's F-16s: Celebrating 40 Years of Combat Airpower." Hill AFB.

Katz, Samuel M. *The Shield of David: The Israel Air Force Into the 1990s*. London: Greenhill Books, 1992.

Koeltzow, Christopher, Brent Peterson, and Eric Williams. "F-16s Unleashed: How They Will Impact Ukraine's War." Center for Strategic & International Studies (CSIS), June 2024.

Kometer, Michael W. *Command in Air War: Centralized versus Decentralized Control of Combat Airpower*. Maxwell AFB, AL: Air University Press, 2007.

Lamb, Michael W. *Operation Allied Force*. Maxwell AFB, AL: Air University Press, 2002.

Lambeth, Benjamin S. *The Transformation of American Air Power*. RAND Corporation, 2000.

Lambeth, Benjamin S. *NATO's Air War for Kosovo: A Strategic and Operational Assessment*. Project Air Force, RAND Corporation, 2001.

Lockheed Martin. *F-16 Fighting Falcon Fast Facts*. September 2024.

North Atlantic Treaty Organization. *The Kosovo Air Campaign: March–June 1999*. Brussels, Belgium: NATO Public Information Office, 2000.

Rosenkranz, Keith. *Vipers in the Storm: Diary of a Gulf War Fighter Pilot*. New York: McGraw-Hill, 2002.

United States Air Force Historical Research Agency / Air Force History Office. *"1999 – Operation Allied Force: Fact Sheet."* US Air Force, n.d.

United States Department of Defense. *Conduct of the Persian Gulf War: Final Report to Congress*. Washington, DC: US Government Printing Office, 1992.

United States Department of Defense. *Kosovo/Operation Allied Force After-Action Report*. Washington, DC: US Government

Printing Office, 2000.

United States Department of Defense. *Gulf War Air Power Survey Summary Report: Vol I-V.* US Government Printing Office, 1993.

Yonay, Ehud. *No Margin for Error: The Making of the Israeli Air Force.* Pantheon Books, 1993.

Secondary Sources:

Atkinson, Rick. *Crusade: The Untold Story of the Persian Gulf War.* New York: Houghton Mifflin, 1993.

Brown, Craig. *Debrief: A Complete History of U.S. Aerial Engagements – 1981 to the Present.* Altgen, PA: Schiffer Publishing, 2007.

Cooper, Tom, and David Nicolle. *Arab MiG-19 & MiG-21 Units in Combat.* Oxford: Osprey Publishing, 2004.

Clodfelter, M. (2006). *Beneficial bombing: The progressive foundations of American air power, 1917–1945.* Lincoln, NE: University of Nebraska Press.

"Futures of the World's Largest F-16 Operators: Why These Six Fleets…" *Military Watch Magazine*, September 2023

Gordon, Michael R., and Bernard E. Trainor. *The Generals' War.* New York: Little, Brown and Company, 1995.

Grier, Peter. "*Package Q: The USAF's First Big Conventional Air Raid of Desert Storm Proved Unexpectedly Dangerous.*" *Air & Space Forces Magazine*, vol. 99, no. 1, Jan. 2016, pp. 60–63. Air & Space Forces Association.

Jane's Information Group. *Jane's Defence Weekly: Coverage of Balkan Air Operations.* London, UK: Jane's Information Group, 1994–2000.

McCarthy, Donald J. *The Sword of David: The Israeli Air Force at War.* New York: Skyhorse Publishing, 2013.

McCarthy, Donald J. *The Raptors: All F-15 and F-16 Aerial Combat Victories*. Altgen, PA: Schiffer Publishing, 2017.

Mehuron, Tamar A. "Operation Desert Storm: Ten Years Later." *Air Force Magazine*, Vol. 84, No. 2, February 2001.

Meilinger, Phillip S. *Air War: Theory and Practice*. Sterling, VA: Brassey's, 1992.

Mersky, Peter. *Israeli Fighter Aces: An Account of the Air Force's Top Pilots*. Specialty Press, 1997.

Michel III, Marshall L. *The Gulf War Air Campaign: Desert Storm 1991*. Osprey Publishing, 2001.

Nordeen, Lon O. *Air Warfare in the Missile Age*. Washington, DC: Smithsonian Books, 1996.

Nordeen, Lon. *Fighters Over Israel*. London: Orion Books, 1990.

Norton, Bill. *Air War on the Edge: A History of the Israel Air Force and its Aircraft Since 1947*. Midland Publishing, 2002.

"Operation Anaconda." *Airspace Historian*, Wordpress.com, 2016.

Rodman, David. *Sword & Shield of Zion: The Israel Air Force in the Arab-Israeli Conflict, 1948-2012*. Liverpool, UK: Liverpool University Press, 2013.

Šafařík, Jan J. *Israeli Air-to-Air Victories: F-16A/B Netz*. Aces.safarikovi.org, n.d.

Šafařík, Jan J. *Israeli Air-to-Air Victories in the Middle East*. Aces.safarikovi.org, n.d.

"SFODA 525: The Battle of Debecka Pass." *SOF News*, 10 Apr. 2017.

Taghvaee, Babak. "Israel's First Air-to-Air Kill Against a Drone." *Combat Aircraft Journal*, 2019.

Tirpak, John A. "Desert Storm's Air Campaign." *Air Force Magazine*, vol. 94, no. 2, Feb. 2011, pp. 36–44.

Ulanoff, Stanley M. and David Eshel. *The Fighting Israeli Air Force*. New York: Arco Publishing, 1985.

www.ingramcontent.com/pod-product-compliance
Lightning Source LLC
Chambersburg PA
CBHW070748200326
41578CB00027B/126